RF Linear Accelerators for Medical and Industrial Applications

For a complete listing of titles in the
Artech House Microwave Library,
turn to the back of this book.

RF Linear Accelerators for Medical and Industrial Applications

Samy Hanna

ARTECH HOUSE

BOSTON | LONDON
artechhouse.com

Library of Congress Cataloging-in-Publication Data
A catalog record for this book is available from the U.S. Library of Congress.

British Library Cataloguing in Publication Data
A catalogue record for this book is available from the British Library.

Cover design by Vicki Kane

ISBN 13: 978-1-60807-090-9

© 2012 ARTECH HOUSE
685 Canton Street
Norwood, MA 02062

All rights reserved. Printed and bound in the United States of America. No part of this book may be reproduced or utilized in any form or by any means, electronic or mechanical, including photocopying, recording, or by any information storage and retrieval system, without permission in writing from the publisher.
  All terms mentioned in this book that are known to be trademarks or service marks have been appropriately capitalized. Artech House cannot attest to the accuracy of this information. Use of a term in this book should not be regarded as affecting the validity of any trademark or service mark.

10 9 8 7 6 5 4 3 2 1

*To my mother, Angelle Hanna,
and to the memory of my
late father, Maurice Hanna*

Contents

1	**Introduction**	1
1.1	Overview of Role of Accelerators in Our Lives	1
1.2	A Glance at the History of Accelerator Development	2
1.2.1	Development of Electrostatic Accelerators	4
1.2.2	Development of Circular Accelerators	6
1.2.3	Development of RF Linear Accelerators	8
1.3	Cancer Radiation Therapy	10
1.4	Market for Industrial Accelerators	11
1.4.1	Producing Medical Radioisotopes for Imaging and Nuclear Medicine	12
1.4.2	Electron Beam Sterilization of Medical Disposables and Food Processing	12
1.4.3	Nondestructive Testing (NDT)	12
1.4.4	Ion Implantation in Semiconductor Chip Fabrication	13
1.4.5	Processing of Polymers	13
1.4.6	Security and Inspection Applications	13
	References	13
2	**Linac Basic Concepts and Constituents**	**15**
2.1	Fundamental Concepts and Definitions	16
2.2	Coupled Accelerator Cavities	21

2.3	Linac's Different Configurations	27
2.3.1	Standing-Wave Linacs	27
2.3.2	TW Linacs	32
2.3.3	Bunching Mechanism	34
2.4	Electron Gun Operation	35
2.5	X-Ray Generation and Target Design	37
2.5.1	Mechanism of Conversion	38
2.5.2	X-Ray Target Design Requirements	38
2.5.3	Target Design Approaches	39
	References	40

3 Linac Supporting System 43

3.1	Introduction—The Linac as a Source for Electron and X-Ray Beams	43
3.2	Linac Auxiliary Systems	45
3.2.1	Linac Vacuum System	45
3.2.2	RF Vacuum Window	46
3.2.3	Linac Water-Cooling System	48
3.3	Radio Frequency (RF) System	49
3.3.1	RF High-Power Sources	50
3.3.2	RF Power Transmission Subsystems	56
	References	60

4 Manufacturing Techniques of Accelerators 61

4.1	Overview of Manufacturing Processes	61
4.2	Material Requirements	63
4.3	Cavity Machining	65
4.4	Chemical Cleaning	65
4.5	Assembly and Bonding Techniques	66
4.5.1	Brazing	66
4.5.2	Diffusion Bonding	70
4.6	Tuning of Linacs	71

4.7	Thermal Outgasing (Bake-Out)	73
4.8	Electron Gun Activation	74
4.9	High-Power RF Conditioning	76
4.10	Linac's Beam Tests and Test Bunkers	78
4.11	Common Manufacturing Issues and Imperfections	80
4.12	Quality Systems in Linac Manufacturing	80
4.12.1	Examples of QC and QA Measures	81
4.12.2	Examples of Statistical Process Control	84
4.13	Guidelines for Linac Buyers and Users	86
	References	88

5 Role of Linear Accelerators in Cancer Radiation Therapy — 91

5.1	Basic Radiation Therapy Concepts and Definitions	91
5.2	Radionuclides-Based Radiation Therapy	95
5.2.1	Brachytherapy	95
5.2.2	Cobalt Teletherapy	95
5.3	Accelerator-Based Radiation Therapy	97
5.4	The Medical Linac Requirements	98
5.5	Clinical Use of Linacs in Radiation Therapy	100
5.5.1	Clinical Requirements	100
5.5.2	Treatment Planning and Simulation	101
5.5.3	Dose Fractionation	101
5.5.4	Rotational Therapy	103
5.6	Conformal Radiation Therapy	104
5.6.1	Multileaf Collimators (MLCs)	105
5.6.2	Intensity-Modulated Radiation Therapy (IMRT)	106
5.6.3	Adaptive Radiation Therapy (ART)	108
5.7	Image-Guided Radiation Therapy (IGRT)	109
5.7.1	Need for IGRT and Treatment Verification	109
5.7.2	Portal Films	110
5.7.3	Electronic Portal Imaging Devices	110

5.7.4	Tomotherapy	112
5.7.5	Robotic Radiosurgery: CyberKnife®	114
5.8	Stereotactic Radiosurgery	115
5.8.1	Gamma Knife™ Radiosurgery	115
5.9	Intraoperative Radiation Therapy	117
5.10	Concluding Remarks	118
	References	119

6 Accelerator-Based Industrial Applications — 121

6.1	Use of Accelerators in Material Processing	121
6.1.1	Electron Beam Current and Energy Requirements	122
6.1.2	Main Applications for Polymer EB Cross-Linking	123
6.1.3	Industrial Material Processing Using X-Ray Radiation	126
6.2	Sterilization of Medical Products and Food Irradiation	127
6.2.1	Medical Product Sterilization	127
6.2.2	Electron Beam Food Processing by Irradiation	129
6.2.3	Mechanism of Killing Pathogens	130
6.3	Environmental Applications of Accelerators	131
6.3.1	Wastewater Treatment	131
6.3.2	EB Treatment of Flue Gases—Reduction of Acid Rain	133
6.4	Nondestructive Testing	135
6.5	Security and Inspection Applications	137
6.5.1	Scanning Units	138
6.5.2	Recent Advancements in Cargo Inspection	140
6.6	Ion Implantation in Semiconductor Chip Fabrication	143
6.6.1	Concept of Operation	143
6.7	Concluding Remarks	144
	References	145

7 Large Accelerators — 149

7.1	Large Accelerator Facilities for High-Energy Physics	149
7.1.1	Linear Colliders	150

7.1.2	Circular Colliders	151
7.1.3	Use of Superconductivity in Large Accelerators	154
7.2	Synchrotrons Sources	157
7.2.1	Synchrotron Storage Rings	157
7.2.2	Wigglers and Undulators	160
7.2.3	Examples of Synchrotron Radiation Applications	161
7.3	Cancer Particle Therapy	163
7.3.1	Advantages of Particle Therapy	164
7.3.2	Particle Therapy Circular Accelerators	166
	References	169

8 Recent Developments and Future Trends in Accelerator Technology — 173

8.1	Free-Electron Laser	174
8.2	Accelerator-Based Neutron Sources	176
8.2.1	Neutron Spallation Sources	176
8.2.2	The Spallation Neutron Source (SNS)	177
8.3	Plasma-Based Accelerators	178
8.3.1	Basic Concept of PWFA	179
8.4	Dielectric Wall Accelerators	180
8.4.1	Basic Concept of DWA	181
8.5	Concluding Remarks	183
	References	184

Glossary — 187

About the Author — 193

Index — 195

1

Introduction

1.1 Overview of Role of Accelerators in Our Lives

When I mention the words *particle accelerators* to my friends and family, they immediately think of high-energy physics experiments with their extremely large machines in national laboratories and research universities. This is not surprising since these accelerators were originally developed for use in "splitting the atom" experiments at the beginning of the last century. It was through these pioneering experiments that we were able to learn about the constituents of the atom. As accelerator technology evolved, we were able to find out more about the basic building blocks of matter and about nature's fundamental forces. Nowadays, a high-energy research particle accelerator can speed up a beam of charged particles to near the speed of light before they collide with a target or with another beam of particles. The sub-nucleus particles created by the collision as well as the radiation emitted are detected and analyzed. Some of these machines accelerate particles in a straight line, and we call them *linear accelerators*, or *linacs* for short. Other high-energy machines are based on rotating the accelerated particles in a circular path, such as the *cyclotrons* and the *synchrotrons*. Actually, one can say that the story of accelerator development goes in step with the progress of high-energy physics. However, along the way, engineers, chemists, and medical physicists have found a number of practical applications using this fascinating technology of particle accelerators.

In fact, the number of accelerators for science experiments (a few hundred) is much smaller than the number of accelerators used commercially to improve and sustain our quality of life. More than 97% of all accelerators are used in commercial applications. There are about 30,000 commercial accelerators quietly running in hospitals, chemical factories, and manufacturing facilities. They impact our lives in many ways. Over the last half century, accelerators have worked behind the scenes to cure thousands of cancer patients around the world, produce isotopes for medical imaging and nuclear medicine, sterilize medical paraphernalia and food products, test hidden pipes and structures for defects and cracks nondestructively, secure our borders, cut and weld steel beams, process semiconductor chips for computers, modify the characteristics of polymers, locate oil and minerals in the earth, and help archaeologists determine the age of ancient artifacts. The current annual global sales of these commercial accelerators are estimated in the billions of dollars.

In this book, we will try to shed light on some of the important applications of RF linear accelerators (linacs), such as the medical applications (Chapter 5) and industrial applications (Chapter 6). In addition to explaining the underlying concepts behind the operation of linacs (Chapter 2), we describe the RF system needed to run them (Chapter 3) and how they are manufactured (Chapter 4). The focus of the book is linacs, however we will mention briefly electrostatic accelerators that are also linear devices. Nevertheless, the word *linac* is reserved for RF linear accelerators. We will also briefly review some of the circular accelerators and sample some of their applications. The brief coverage in this chapter of the electrostatic and circular accelerators will help us to have a better understanding of the industrial and scientific applications of accelerators in the last three chapters of this book.

In this introductory chapter we will very briefly look at the history of accelerating development with their different types (Section 1.2). We will highlight one of the major medical applications of accelerators, namely cancer radiation therapy (Section 1.3). In the last section of this chapter we list some of the industrial applications of accelerators.

1.2 A Glance at the History of Accelerator Development

The driving force behind the development of accelerators has been directly tied to the advancement of high-energy physics research [1–3]. In 1919, Ernest Rutherford, a renowned physicist, successfully induced nuclear reactions using natural radioactive materials. This groundbreaking work motivated scientists to develop devices that would eventually accelerate particles to voltages higher than those obtained from radioactive materials and would allow for careful and controlled study of the atom nucleus. Accelerator development continued for nine more decades. In this section, we will briefly highlight the significant milestones of this development. Following the path that accelerator development took, we find that most of the early development approaches focused on the simpler concept of using a dc voltage to provide the electric field required for accelerating the charged particles. This is the electrostatic approach that resulted in today's *electrostatic accelerators*. On the other hand, time-varying voltages were proposed almost at the same time, though not realized right away, and eventually they resulted in two groups of accelerators. The first is the group of circular accelerators, namely the *cyclotron*, the *betatron*, and the *synchrotron*. The second is the *RF linear accelerator*, which is the focus of this book. We map the course of early accelerator development in Figure 1.1.

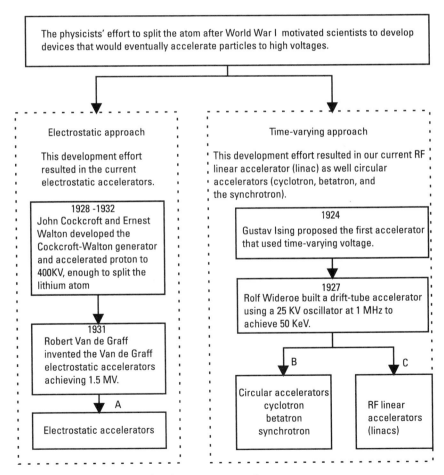

Figure 1.1 Early accelerator development progression.

1.2.1 Development of Electrostatic Accelerators

In 1932, two of Ernest Rutherford's students, John D. Cockcroft and Ernest T. S. Walton, designed and built an electrostatic accelerator at the Cavendish in Cambridge [1, 2]. They built a voltage multiplier consisting of a set of capacitors connected to rectifying diodes acting as switches. By opening and closing these switches in a certain sequence, they could achieve a potential of 400 kV from their 200-kV transformer. With this device they were able to accelerate protons down a vacuum tube eight feet long towards a lithium target. This was sufficient to disintegrate the nuclei of the lithium atom.

The original *Cockcroft-Walton accelerator* is now kept in the Science Museum in London and shown schematically in Figure 1.2 (from [2], with modifications).

Most of high-energy electrostatic accelerators that we know today are based on the invention of Robert James Van de Graff and thus is known now as the Van de Graff generator [4]. The concept behind this accelerator is depicted in Figure 1.3. His idea was to build a machine that would transport charges mechanically. The charges are then delivered to an upper terminal where they accumulate. The buildup of charges translates to a buildup of voltage on that terminal, which has the shape of a sphere. This shape allows for maximum buildup of charges without creating sparks. The resulting built-up voltage is then used to accelerate particles from that high voltage to ground potential across an evacuated accelerating channel.

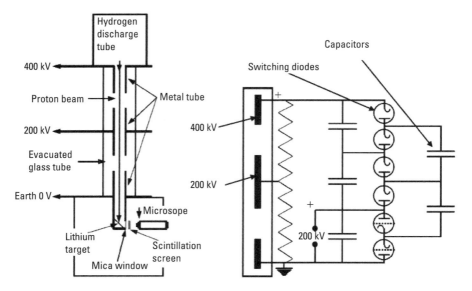

Figure 1.2 Cockcroft and Walton setup. (a) the accelerator, and (b) the dc voltage multiplier.

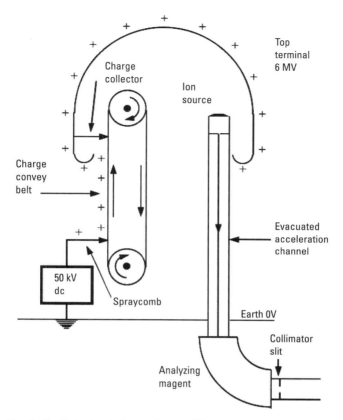

Figure 1.3 Van de Graff electrostatic accelerator [2].

Ray Herb [5] added three key improvements to this machine. First, he replaced the vacuum with pressurized gas that would stand higher voltages. Second, he changed the belt carrying the charges from a continuous smooth belt to one with steel cylinders held apart by nylon insulating links in the form of a chain. A third contribution was to put conducting rings along the accelerating channel, which helped in shaping the accelerating field more precisely. These improvements allowed the Van de Graff electrostatic accelerator lend itself to more practical and commercial applications, such as ion implantation of semiconductor wafers as discussed in Chapter 6.

Electrostatic accelerators are limited in the maximum energy they can deliver to a charged particle. This final energy is basically the product of the charge times the dc potential difference that can be maintained. In practice this potential difference is limited by electric breakdown to no more than a few tens of megavolts. Time-varying accelerators circumvent this limitation, as we will see below, in both circular and RF linear accelerators.

1.2.2 Development of Circular Accelerators

Ernest Lawrence invented the first circular accelerator, the *cyclotron* [6]. He relied on the fact that a moving charge bends in a curved trajectory under the effect of a dc magnetic field applied normally to the plane of motion of the charged particle beam. An RF driving voltage is applied across a gap between two D-shaped half cylinders as shown in Figure 1.4. As the charged particles gain energy every time they cross the gap between the "Dees," they spiral outwards in an increasing radius. The two effects—of increase in radius and the gain in energy (speed)—compensate for each other, and thus the time for each revolution stays constant. If the applied magnetic field is chosen correctly, the particle circulation frequency will be in synchronism with the RF oscillating field across the gap.

The operation can be described in a few simple equations as follows.

Recalling that a particle of mass m traveling in a circular orbit of radius r with velocity v will be subject to a centrifugal force F given by

$$F = mv/r^2 \qquad (1.1)$$

Now, in the presence of a magnetic field B perpendicular to the plane of motion of the particle with charge q and moving with velocity v, the particle would experience a compensating force toward the center of the circle of revolution of

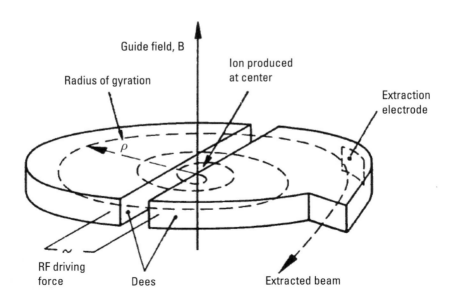

Figure 1.4 Concept of operation of the cyclotron [2].

$$F = qvB \tag{1.2}$$

By equating the above two equations, we see that

$$\omega = v/r = qmB \tag{1.3}$$

From (1.3), we get the important notion that for a particle of charge q and mass m in a field B perpendicular the plane of motion of the particle, *the frequency of rotation ω is independent of the radius of the orbit r.*

From the beginning, it was clear that the cyclotron would not be suitable for accelerating electrons that would move at a speed close to that of light once they reach energies close to a million volts of acceleration and therefore loose synchronism with the accelerating voltage. This was not the case for another circular accelerator, the *betatron*, which Donald Krest developed in 1940 [7]. The basic concept of its operation is depicted in Figure 1.5.

A basic relation between a time-varying magnetic field and the resulting electric field is Faraday's law. The simplest manifestation of this law is the observation that changes in the magnetic environment of a loop of wire will cause a voltage to be "induced" in the loop. In the configuration shown in Figure 1.5 [2], as the magnetic field B_z changes, an orbiting electron (analogous to a current in a loop of wire) experiences an "induced" azimuthal tangential electric field E_s, which drives the electron to higher and higher energy. Actually, the concept of operation of the betatron is analogous to the electrical transformer, both of which are based on Faraday's law. In the betatron, the primary part of the transformer is the electromagnet providing the time-varying magnetic field and the secondary part is the orbiting electron. The betatron configuration is shown schematically in Figure 1.6 [2].

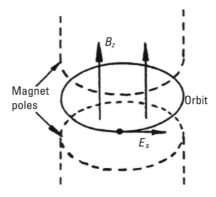

Figure 1.5   Basic concept of the betatron.

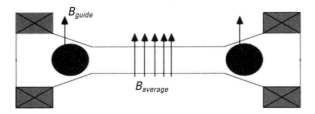

Figure 1.6 Schematic for the betatron.

The magnetic field in the betarton does both bending (B-guide) and acceleration (B-average) of the particle (usually electrons), provided that the bending magnetic field is increasing with time. The increase in the magnetic field strength matches the increase in particle energy, so that the orbit radius is kept nearly constant, and the induced electric field provides acceleration of the particles. The magnet must be carefully designed so that the field strength at the orbit radius (B- guide) is equal to half the average field strength (B-average) linking the orbit [7].

McMillan [8] and Veksler [9] independently invented the third circular accelerator after World War II. In the *synchrotron*, the particles are constrained to circulate in a constant radius in a ring under vacuum. The magnetic guiding field increases with the particle energy, so as to keep the orbit at the same radius. The accelerating force is supplied by azimuthal tangential electric field provided by RF source (cavity); see Figure 1.7. Once the particle beam reaches the targeted energy, the RF fields are there to make up for the beam's energy loss due to radiation from the circulating beam. The radiated energy is based on the fact that deflecting electron beams in a magnetic field emits electromagnetic waves. This *synchrotron radiation* was soon put to use for both medical and industrial applications. We will discuss the operation of the synchrotron briefly in Chapter 7 when we discuss proton therapy machines for treating cancer and storage ring colliders, including the most recent proton synchrotron at the European Organization for Nuclear Research (CERN) near Geneva, Switzerland—namely, the Large Hadron Collider (LHC).

1.2.3 Development of RF Linear Accelerators

In 1924 Gustaf Ising, a Swedish physicist, proposed accelerating charged particles using alternating electric fields [1, 2]. His proposed accelerator configuration consisted of a straight glass tube under vacuum that housed a sequence of metallic tubes with holes for the beam to drift through them, and hence these tubes are called "drift tubes" and this accelerator structure is called a *drift-tube accelerator*. The particles are accelerated by the field existing in the gaps between the drift tubes, as shown in Figure 1.8. The accelerating field was established by

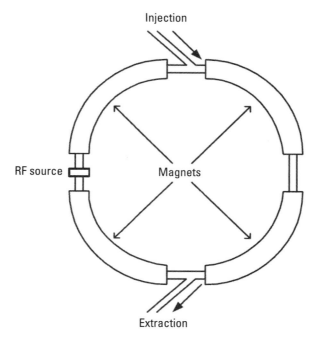

Figure 1.7 Basic principle of the synchrotron.

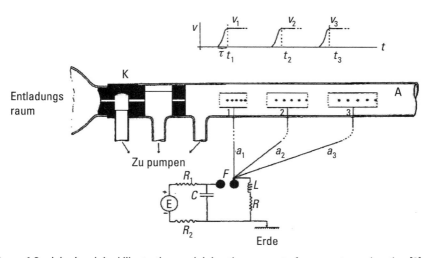

Figure 1.8 Ising's original illustration explaining the concept of resonant acceleration [2].

the pulsed voltage applied across adjacent drift tubes and were generated by a capacitor discharge at electrode F, in Figure 1.8. In order for the applied voltage to be synchronized with the particles moving downstream, the transmission lines a_1, a_2, a_3 were chosen with different lengths to add the proper delay from the voltage source to each of the drift tubes.

It is to be noted that although Ising used a series of timed dc pulses, he could as well have used an RF (radio frequency) oscillator to provide the voltage across successive gaps. A young Norwegian engineering graduate student, Rolf Wideröe, who was earning his Ph.D. in Aachen, Germany, in 1927, realized this fact. An alternating voltage powered his linear accelerator, and thus he is considered the inventor of the world's first RF linear accelerator (linac). His accelerator propelled potassium ions through an 88-cm-long glass tube, achieving an energy gain equivalent to twice the peak voltage he used. The basic concept of the Wideröe accelerator is depicted in Figure 1.9.

In order for the Wideröe linac to accelerate electrons to velocities close to the speed of light, the accelerator had to be either very long, or the wavelength of the driving RF voltage had to be short. Thus, practical uses of this type of accelerator had to wait until after World War II to benefit from the advancement in microwave sources. Since that time, there have been several variants of accelerating structure design, although the underlying principle remains unchanged. We will discuss the concept of operation of RF linear accelerators with their different design configurations in the next chapter.

In the remaining two sections of this chapter we list some of the applications of accelerators, including the medical and industrial applications; both are discussed in more detail in other chapters of this book.

1.3 Cancer Radiation Therapy

Cancer is a major public health problem in many parts of the world. In the United States, cancer is the second leading cause of death (more than half a million deaths in 2009 were due to cancer) [10]. Fortunately, cancer death rates have been decreasing lately. New advances in the diagnosis, screening,

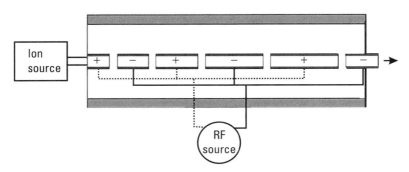

Figure 1.9 Basic concept of the Wideröe RF linear accelerator.

and treatment of cancer in the last few decades have contributed effectively to the discovery and curing of this disease. The most common cancer treatment modalities include surgery, chemotherapy, and radiation therapy.

Radiation therapy (RT), also known as *radiotherapy,* is an effective method of treatment for many different forms of cancer using ionizing radiation. About half of all cancer patients receive radiation therapy, either alone or in combination with one or two of the above-mentioned treatment modalities. The ionizing radiation in this type of cancer treatment is produced from machines based on an electron linear accelerator (linac) or from a radioactive material. The goal of most radiation therapies is to destroy cancerous tissue. There are basically two types of RT, namely, *brachytherapy* and *external radiation therapy.* Brachytherapy is a type of radiation therapy in which radioactive materials are placed close to or inside the patient's body. External radiation therapy is any type of RT in which the radiation source is located at a distance from the patient's body. By referring to radiation therapy or RT in the remaining pages of the book, we mean the externally applied radiation produced by accelerator-based machines, since this is the focus of this book. In this medical application, more than 10,000 linacs are used around the world. Cancer radiation therapy based on linacs will be discussed in more detail in Chapter 5. Radiation therapy facilities using protons accelerated by synchrotron accelerators are covered in Chapter 7.

One challenging objective of RT is to maximize the radiation dose to the tumor while minimizing the effects of radiation on the surrounding healthy tissue. This is most critical when the tumor's surroundings include sensitive tissues such as eyes, spinal cord, or lungs. Achieving this objective is very dependent on the accuracy and precision of the process of dose delivery. In Chapter 5, we will discuss the requirement for RT machines and type of linacs used, as well as new advances in beam delivery.

1.4 Market for Industrial Accelerators

The size of the market for industrial accelerators more than matches that for the radiation therapy discussed briefly above. It is estimated that the number of industrial accelerators sold so far is more than 18,000, with annual sales exceeding 900 units. This translates into an annual market of $2 billion [11]. Originally, the designs of many of the accelerators used in industrial applications were developed in research laboratories for physics experiments. However, once the technology was proven, these accelerators were put to use for different applications. In Chapter 6, we discuss in more detail some of examples of these applications. Let us briefly look at some of these applications.

1.4.1 Producing Medical Radioisotopes for Imaging and Nuclear Medicine

Over 10,000 hospitals worldwide use radioisotopes in medicine with about 90% of the procedures for diagnosis and about 10% for treatments. In the United States, there are some 18 million nuclear medicine procedures done per year using radioisotopes [12]. The short-lived isotopes are used as radioactive tracers that emit gamma rays from within the body, permitting specific physiological processes to be scrutinized to provide information about the functioning of a person's specific organs or to treat certain diseases. It is expected that the demand for nuclear medicine procedures using radioisotopes will continue to increase, especially in developing countries. Artificially made radioisotopes, including those for medical use, are mainly produced by research reactors. Some facilities producing isotopes utilize particle accelerators, mostly circular accelerators (cyclotrons) and some linear accelerators (linacs).

1.4.2 Electron Beam Sterilization of Medical Disposables and Food Processing

Although electron beam (EB) sterilization processing of medical devices was used in the 1950s, it was not widely used until the 1970s, when the accelerators used became more efficient and reliable. In this application, accelerators produce high-energy electrons. The high-energy electron beam penetrates through the packaging and destroys the DNA chains of the microorganisms, thus sterilizing the product in its final packaging. This sterilizing technique is the approach of choice for many medical disposable manufacturers, offering the lowest cost of sterilization for large capacities.

In many countries, high-energy electron and X-ray beams are used to irradiate food. The purpose of radiation preservation of food is to prevent undesirable changes in the food during shipment and for longer shelf life. In addition to preservation, radiation processing can also provide a means of disinfecting, by destroying any pathogens present.

1.4.3 Nondestructive Testing (NDT)

Standard X-ray tubes have been used for several decades to inspect thin objects to detect hidden defects or flaws. However, they are typically effective only for the inspection of thin articles. For thick objects, higher energy X-ray beams are needed for effective radiographic inspection. For example, a 250 mm–thick of steel plate or a 1 meter–thick concrete wall can be inspected with the help of high-energy X-ray beams. These are normally produced by electron linacs using tungsten targeted to produce X-rays capable of penetrating thick objects.

1.4.4 Ion Implantation in Semiconductor Chip Fabrication

Ion implanters are used in the integrated-circuit manufacturing industry. The ion beam implanter function is to alter the near-surface properties of semiconductor materials by implanting dopant ions that change the electrical characteristic of the semiconductor. A typical ion implanter used in the manufacture of electronic devices involves generating and accelerating an ion beam and steering it into the semiconductor substrate so that the ions come to rest beneath the surface at a specified depth. Among semiconductor-processing techniques, ion implantation is nearly unique in that process parameters, such as concentration and depth of the desired dopant, are specified directly in the equipment settings for controlling the implant dose and energy, respectively. The progress made in electrostatic accelerators in the 1960s made the Van de Graff electrostatic accelerator a practical reliable and source of ions for the semiconductor industry as discussed in Chapter 6 of this book.

1.4.5 Processing of Polymers

The radiation source for processing polymers may be an electron beam accelerator or a radioactive source, such as Cobalt-60. However, EB accelerators are more commonly used for commercial radiation modification of polymers. Their dose rate is much higher, and they allow more efficient movement of the product in front of the radiation source. Thus, it is generally a more cost-effective way of delivering dose to the product. The energy of the electron beam determines its depth of penetration into the product.

1.4.6 Security and Inspection Applications

In many of the world's ports we can now find high-energy X-ray linear accelerators. These accelerators can reliably generate the energies required to penetrate loaded containers while retaining the sensitivity to discriminate among different materials and to provide high quality images for cargo inspection requirements. In the last decade multiple countries have started R&D programs at their research centers and universities to develop systems for border security, airline-cargo inspection, and first response in the investigation of unknown packages.

References

[1] Sessler, A., and E. Wilson, *Engines of Discovery: A Century of Particle Accelerators*, World Scientific Publishing Co., Singapore, 2007.

[2] Bryant, P. J. "A Brief History and Review of Accelerators," *CERN Accelerator School Fifth General Accelerator Physics Course*, Jyväskylä, Finland, September 7–18, 1992, 1–16.

[3] Wilson, E. J. N., "History of Accelerators," *An Introduction to Particle Accelerators*, Oxford University Press, 2001.

[4] Van de Graaf, R. J., "A 1,500,000 volt Electrostatic Generator," *Phys. Rev.*, Vol. 387, 1931, pp. 1919–1920; Courant, E. D., "Early Milestones in the Evolution of Accelerators," *Rev. of Accelerator Science and Technology*, Vol. 1, 2008, pp 1–5.

[5] Herb, R. G., "Early Electrostatic Accelerators and Some Later Developments," *IEEE Transactions*, NS-30, No. 2, 1983, pp. 1359–1362.

[6] Lawrence, E. O., and M. S. Livingstone, "The Production of High Speed Light Ions Without the Use of High Voltages," *Phys. Rev.*, Vol. 40, 1932, pp. 19–35.

[7] Krest, D. W., "Induction Electron Accelerator," *Phys. Rev.*, Vol 59, 1940, p. 110.

[8] McMillan, E. M., "The Synchrotron—A Proposed High Energy Particle Accelerator," *Phys. Rev.*, Vol. 68, 1945, pp. 153–158.

[9] Veksler, V. I., "A New Method of Acceleration of Relativistic Particles," *Journal of Physics UdSSR*, Vol. 9, 1945, pp. 153–158.

[10] Data from the Centers for Disease Control and Prevention (CDC), http://www.cdc.gov/nchs/fastats/lcod.htm, last modified June 28, 2010, accessed January 3, 2011.

[11] R. W. Hamm, "Industrial Accelerators," *Reviews of Accelerator Science and Technology*, Vol.1 (2008), World Scientific Publishing Co., pp 163–184.

[12] World Nuclear Association, "Radioisotopes in Medicine," www.world-nuclear.org/info/inf55.htm, last modified February 2010, accessed March 2010.

2

Linac Basic Concepts and Constituents

In its simplest definition, an RF linear accelerator (linac) is a device that uses electromagnetic waves, in the microwave range, to accelerate charged particles such as electrons. Some of the medical and industrial applications employ the resulting accelerated high-energy particle beams. In many other applications, the final useful output of the linac is a beam of high-energy X-rays resulting from the collision of the accelerated electron beam with heavy metal target. Figure 2.1 shows a simplified functional diagram of a linac that receives electrons from an electron gun and accelerates them to produce high-energy X-rays. This beam is accelerated in a series of accelerating cavities by forces exerted on the electrons by the fields of a microwave, which is fed from an external microwave source such as a klystron or a magnetron.

We will discuss different types of accelerators used in medical and industrial applications in Chapters 5 and 6, respectively. In this chapter, I will guide you in developing an understanding of the fundamental concepts of operation of these accelerators. Formal analysis of linacs would require solving Maxwell's equations governing the electromagnetic fields in the linac subjected to their boundary conditions. In most cases computer simulation is a must for any design of a linac. Here, I rely on describing the concepts behind the physics with the help of explanatory illustrations and a few basic formulae, as needed. Readers wanting to learn about linacs in more depth can consult one of the accelerators books [1–3].

I begin by using the simplest accelerator arrangements, such as the dc two-electrode and the RF single-cavity accelerating configurations, to introduce the essential definitions and fundamental concepts we need. Most of the RF linacs can be viewed as a chain of coupled cavities. To understand the underlying

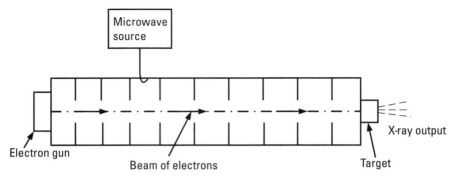

Figure 2.1 A simplified electron linac.

physics of the linac operation, I employ the coupled-circuit model to establish many of the linac operational concepts and parameters. Also in this chapter, I discuss other linac constituents such as the electron gun and the X-ray target.

2.1 Fundamental Concepts and Definitions

Let us begin first by reviewing some of the simple concepts and definitions that we are going to need. I will start with the simplest configuration one can use to accelerate charged particles. It is comprised of two plates (electrodes) with a dc voltage, V_a, applied between them, as shown in Figure 2.2.

A particle with a positive charge q existing in the space between the two electrodes would experience a force F_q. This force is proportion to the charge

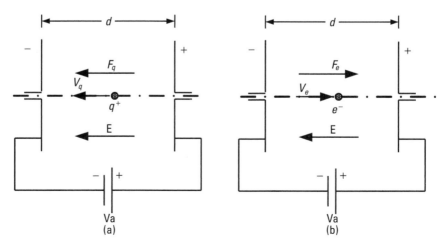

Figure 2.2 A simple two-plate accelerator (a) acceleration of a positively charged particle, and (b) acceleration of an electron.

q and the *accelerator voltage* V_a and inversely proportional to d, the distance between the electrodes.

$$F_q = q \frac{(V_a)}{d} \quad (2.1)$$

The term $\frac{(V_a)}{d}$ in (2.1) is called the *electric field E* and its strength is measured in volts per meter (V/m). We sense the presence of an electric field by the force it exerts on a charged particle. The direction of the electric field, as indicated in Figure 2.2, is marked by an arrow pointing in the direction in which a positively charged particle would move in the presence of this field.

If the charged particle is starting at rest at one plate, then the accelerated particle would leave this simple accelerator on the other side with *output energy* U_o:

$$U_o = qV_a \quad (2.2)$$

If the accelerated particle is an electron, as a negatively charged particle it would then move opposite to the direction of E (see Figure 2.2(b)), and our electron in this simple electron accelerator would be accelerated from rest to an output energy of U_0

$$U_o = -eV_a \quad (2.3)$$

where e is the charge of the electron 1.6×10^{-19} C (coulomb). For this reason it is common to express the energy of accelerated particles (electrons, protons, ions, etc.) using the unit of *electron-volt* (eV). Thus, an electron-volt is the energy an electron gains when accelerated by an accelerating voltage of one volt. In most of the applications we cover in this book, the energy is measured in millions of electron-volts; so the energy unit commonly used is the megaelectron-volt (1 MeV = 1×10^6 eV). In many high-energy physics research accelerators the energy is measured in billions of electron-volts; so the energy unit used is the gigaelectron-volt (1 GeV = 1×10^9 eV).

Now, let us imagine that the dc voltage applied between the electrodes would be one million volts; then it is more fitting to measure the electric field strength in M V/m, megavolts/meter. Actually, our "virtual" accelerator would not work if the distance between the electrodes is, say, less than 0.3m. The reason is that, although air is normally a good insulator, air begins to break down at electric field strength of about 3 MV/m. So let us enclose our simple two-plate accelerator and evacuate that enclosure to avoid breakdown. Another benefit for keeping the accelerator under good vacuum is to minimize the scattering of the accelerated charged particles by reducing the chances of collision between the particles being accelerated and any residual gas molecules in the path of the accelerated particle.

We can even move closer to our practical linac by replacing our dc voltage source by an alternating voltage source, $V(t)$. This will result in a time-varying electric field $E(t)$. We then will notice that, as we increase the frequency of our source towards the radio frequency (RF) range, we lose a lot of energy that would radiate away. So let us enclose our electron in a metallic *cavity*, which we can evacuate to keep under good vacuum (Figure 2.3).

If an electromagnetic wave is launched inside such a conducting cavity, its interior metallic surfaces reflect the wave back and forth. When a wave having a specific frequency, called *resonant frequency*, that is resonant with the cavity enters, it bounces within the cavity with low loss. As more wave energy enters the cavity, it combines with and reinforces the bouncing wave, increasing its intensity. Such resonant cavities can be used to efficiently accelerate a charged particle, provided that the particle is inserted in the cavity at a time when the electric field in the cavity is at or near its peak. Actually, the resonant cavity is the most fundamental element in a linac and constitutes the building block for the majority of RF accelerators.

Now, let us look at the parameters we use to characterize an accelerating cavity. The first one is the resonant frequency f_0. In order for our cavity to be a resonator, the dimensions of a cavity should be of the order of the RF wavelength λ_{RF}, which is related to the RF frequency f_{RF} by the speed of light c,

$$c = f_{RF} \lambda_{RF} \tag{2.4}$$

A common approach to analyze an accelerating resonant cavity is to solve Maxwell's equations governing the electromagnetic fields in the cavity and subject to its boundary conditions. However, to keep our analysis less dependent on differential equations, we use here the equivalent circuit approach in

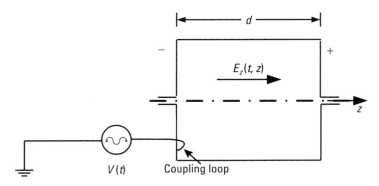

Figure 2.3 A simple RF cavity.

modeling our resonant cavity [2, 5]. The equivalent circuit is shown in Figure 2.4. The cavity is modeled as a parallel RLC resonant circuit driven by a time-varying source.

The inductance L models the magnetic properties of the cavity when hosting the electromagnetic fields. The capacitance C is a representation of the electric properties of the cavity. The resistance R models the losses due to *RF* power dissipated in the cavity walls. Let us now consider some of the parameters commonly used to characterize such a resonant circuit. The resonant frequency is expressed in terms of the circuit elements, its inductance L and its capacitance C (assuming the resistance R is small), as given in (2.5),

$$f_0 = \frac{1}{2\pi\sqrt{LC}} \tag{2.5}$$

The stored energy U is given by

$$U = \frac{1}{2}CV_0^2 \tag{2.6}$$

where V_0 is the peak voltage across the parallel circuit. The condition of resonance in our RLC circuit occurs at the frequency at which the energy stored in the inductance in one half of the RF cycle is equal to the energy stored in the capacitance in the other half of the cycle. Similarly, our RF cavity would resonate when the energy stored in the electric field in half of the RF cycle is equal to the energy stored in the magnetic field in the other half of the RF cycle. So we can express the energy stored in a cavity in terms of its electric field E or its magnetic field H by integrating each of these fields over the volume of the cavity v as demonstrated in (2.7).

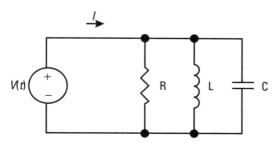

Figure 2.4 Simple RLC resonant circuit.

$$U = \frac{1}{2}\mu_0 \int_v |H|^2 \, dv = \frac{1}{2}\varepsilon_0 \int_v |E|^2 \, dv \qquad (2.7)$$

Where μ_0 and ε_0 are the free-space permeability and permittivity, respectively.

Going back to our simple circuit, the average power dissipated in the circuit is given by the power dissipated in its parallel resistance R:

$$P = \frac{1}{2}\frac{V_0^2}{R} \qquad (2.8)$$

In the RF cavity, power loss is proportional to the strength of the magnetic field at the cavity walls and the surface resistance of the cavity. For this reason, it has been established that oxygen-free high-conductivity (OFHC) copper is the preferred material for room temperature RF cavities for its very low resistivity. In order to complete the analogy between the RF cavity and its RLC model, let us define a parameter that would correspond to the resistance R in the circuit. This parameter is called the cavity shunt impedance R_{sh}. It determines how much acceleration one gets from the cavity for a given wall dissipation.

$$R_{sh} = \frac{1}{2}\frac{V_{acc}}{P_{cav}} \qquad (2.9)$$

It is to be noted that sometimes the shunt impedance is defined without the factor of 2 in the denominator, as in (2.9) above, so be careful. From the definition of R_{sh} we see that in order to maximize acceleration for given cavity loss P_{cav}, one must maximize the shunt impedance. This is done by the choice of a low-loss material for the cavity walls and by optimizing the cavity design, since R_{sh} depends on the cavity material and its geometry.

An important parameter for any resonator is the ratio of the stored energy in it to the power dissipated in it in one cycle. This important parameter is called the *quality factor* Q, which is given by:

$$Q = 2\pi f_0 \left(\frac{U}{P}\right) \qquad (2.10)$$

A typical value of Q for a normal conducting copper cavity is of the order of $Q_{Cu} = 10^4$ (note that a superconducting cavity can have the quality factor of about $Q_{sc} = 10^{10}$).

Even if we use a good material for our cavity (a high Q) and a good design (a high R_{sh}), we are limited in the maximum acceleration we can deliver to the particle being accelerated using a single cavity. The limitation comes from the highest electric field that can be sustained in the cavity without arcing (even if we maintain good vacuum in the cavity). The obvious solution then is to use multiple successive cavities so the accelerated particle, such as an electron, can gain more energy as it passes from one cavity to the next. However, we need to make sure that our electron would be in the center of each cavity when the electric field in each cavity has the right polarity to keep the electron accelerated in the direction it is traveling. We then have to synchronize all the RF sources that energize the individual cavities. A better approach is to feed all cavities from one source and connect, or *couple*, these cavities together.

2.2 Coupled Accelerator Cavities

Let us consider first the simple example of two coupled cavities shown in Figure 2.5. An analogous mechanical system would be two pendulums connected with a weak spring (see Figure 2.6.) This mechanical system can oscillate in two different ways (modes) with two different frequencies. One mode of oscillation is when pendulums are swinging "in phase," (Figure 2.6a). We call this the *zero mode*. The other mode is when they swing in opposite directions (assuming that they do not collide); we call this the π *mode* (Figure 2.6b). The same effect exists in a two-cavity system: fields in adjacent cavities can have the same (Figure 2.5a) or opposite directions (Figure 2.5b). Similarly, these two modes are called *zero mode* and *π-mode*, corresponding to the phase shift between fields in neighboring cavities. Difference in frequencies of the two modes is larger if the coupling (the spring) between two cavities (pendulums) is stronger. The strength of the coupling between the two cavities is represented as a dimensionless factor k.

The next step is to analyze a chain of three coupled cavities (see Figure 2.7). Here we also represent the strength of the coupling between every two adjacent cavities by the dimensionless factor k.

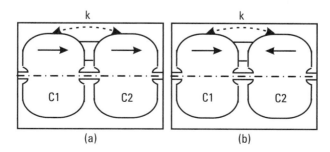

Figure 2.5 Two coupled RF cavities: (a) zero mode oscillation, and (b) π mode oscillation.

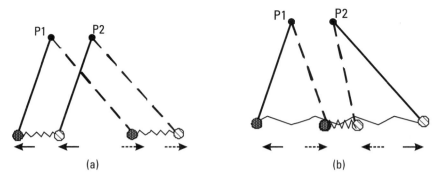

Figure 2.6 Two coupled pendulums: (a) zero mode oscillation, and (b) π mode oscillation.

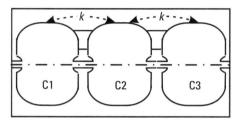

Figure 2.7 A chain of three coupled cavities.

The coupled cavities are usually analyzed by using the equivalent circuit representation of a chain of resonant RLC circuits magnetically coupled [2, 5]. When the results of the analysis are applied to our three coupled cavities, we get three modes; each one has a specific RF field configuration and a corresponding frequency. If we assume that the three coupled cavities are identical and each has an uncoupled frequency f_0 and normalized field of amplitude of unity, then the resulting three modes, in order of increasing frequency, are:

The zero mode. It is called the 0 mode because all cavities have zero relative phase difference between each other.
The frequency of this mode is: $f_0 = \dfrac{f_0}{\sqrt{1+k}}$, and relative field $E_0 = \begin{bmatrix} 1 \\ 1 \\ 1 \end{bmatrix}$

The $\pi/2$ mode. In this mode, the relative phase difference between each cavity and its neighbor is $\pi/2$.
The frequency of this mode is: $f_{\pi/2} = f_0$, and relative field $E_{\pi/2} = \begin{bmatrix} 1 \\ 0 \\ -1 \end{bmatrix}$

The π mode. In this mode, the relative phase difference between each cavity and its neighbor is π.

The frequency of this mode is: $f_\pi = \dfrac{f_0}{\sqrt{1-k}}$, and relative field $E_\pi = \begin{bmatrix} 1 \\ -1 \\ 1 \end{bmatrix}$

We notice that we can deduce the coupling factor k in terms of the lowest frequency f_0 (the zero mode frequency), the highest frequency f_π (the π-mode frequency), and the middle frequency (the $\pi/2$-mode frequency). Equation (2.11), below, gives this formula (for $k \ll 1$, which is the case in many of the practical cases).

$$k \approx \frac{f_\pi - f_0}{f_{\frac{\pi}{2}}} \tag{2.11}$$

Each of the above natural modes of oscillation represents a possible steady-state operation for the three-cavity chain where each mode has a resonant frequency and a corresponding field pattern with a particular phase shift from one cavity to next. Figure 2.8 gives simplified presentations of field orientations for the three modes. By choosing the excitation frequency of an external RF source, we can decide in which mode to operate this chain. For example, if the exciting frequency matches the zero mode, , then the three-cavity chain would

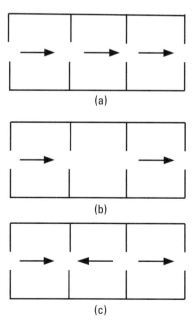

Figure 2.8 Presentations of field orientations for the three modes: (a) zero mode, (b) $\pi/2$ mode, and (c) π mode.

operate in the zero mode and the fields in the three coupled cavities would have the same amplitude and they would all be "in phase." If we excite the π mode, we will have the field in each cavity of the chain 180° "out of phase" form the adjacent cavity. We will discuss this mode further since it is usually the working mode for superconducting accelerators. It is interesting to note that if the $\pi/2$ mode is excited, then the middle cavity in this chain is not excited (i.e., having zero field). We will learn more about the unique characteristics of the $\pi/2$ mode when we discuss some linacs that operate at this mode, such as those used for cancer radiation therapy.

Let us now extend our analysis to a chain of seven coupled cavities. Similar to the above analysis, solving we get seven modes of oscillation. In Figure 2.9, we only show the presentations of field orientations for the mode with the lowest frequency (0 mode) and the highest frequency mode (π mode). It is to be noted that the arrows are just a presentation of the directions of the electric filed at one instant, but these fields are varying sinusoidally with time. In Figure 2.10, we plot frequencies of all seven modes versus the cavity-to-cavity phase shift for each mode. The resulting graph is a cosine-like curve and is called the *dispersion curve*, where the first point represents the 0 mode and the seventh point corresponds to the π mode. The seven modes would be: 0 mode, $\pi/6$ mode, $\pi/3$ mode, $\pi/2$ mode, $2\pi/3$ mode, $5\pi/6$ mode, and π mode. Similarly, a chain of N-coupled cavities will always have N-modes of oscillation.

Before we go further, let us pause here and look at the mechanisms of coupling between cavities. There are two ways to couple a chain of cavities together:

Electric coupling: The cavities are coupled by the electric field shared by every two adjacent cavities through the coupling apertures along the common axis (Figure 2.11).

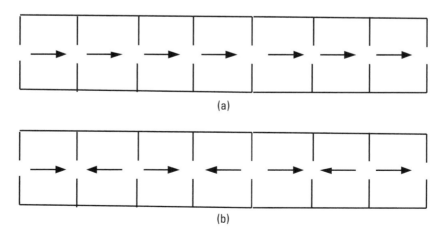

Figure 2.9 A chain of seven coupled RF cavities: (a) 0 mode, and (b) π mode.

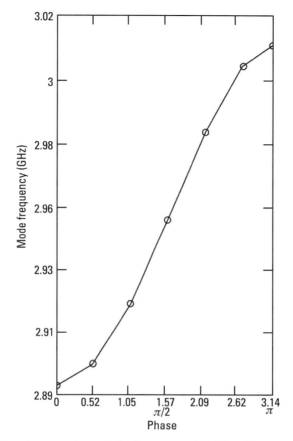

Figure 2.10 Dispersion curve for a chain of seven coupled RF cavities.

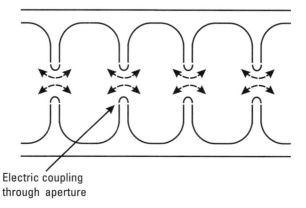

Figure 2.11 Mechanisms of coupling between cavities (electrical coupling).

Magnetic coupling: Here we have to open slots or apertures in the common wall between every two adjacent cavities in the regions of high magnetic field. The cavities are coupled through the magnetic flux linking both adjacent cavities (Figure 2.12).

Coupled cavities are used in accelerators to host the electromagnetic fields needed to accelerate charged particles. The electric field component is responsible for exerting the accelerating force on the particles being accelerated. We can make our accelerator more efficient by concentrating the electric field on an axis by introducing *nose cones* as shown in Figure 2.13. Another improvement would be to make the opening for the beam as small as possible. This later change would limit the coupling through the axial bore. For this reason, many

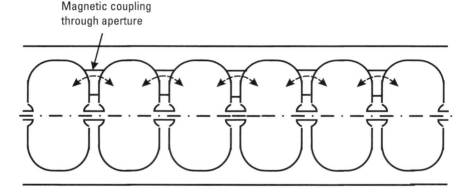

Figure 2.12 Mechanisms of coupling between cavities (magnetic coupling).

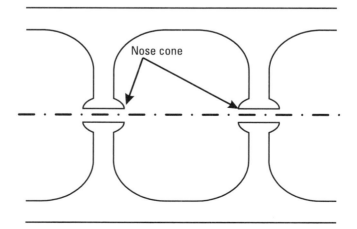

Figure 2.13 Incorporating nose cones to enhance the linac efficiency.

accelerators designs use off-axis apertures to connect adjacent cavities, such as those shown in Figure 2.12.

2.3 Linac's Different Configurations

The function of the linear accelerator, or linac, is to host electromagnetic waves that efficiently deliver energy to the charged particles being accelerated. There are two conditions that have to be satisfied in the linac structure for it to be successful in achieving the above function. These are:

1. The electromagnetic waves in the linac should have an electric field component along the direction of motion of the accelerated particle.
2. The particle and the wave must move in synchronism. For this reason the particles are first bunched together upon entrance into the linac. Then the "bunch" of particles can move in synchronism with the accelerating fields.

There are two ways to describe the operation of a linac. In the first, we analyze its operation as a chain of coupled resonant cavities operating at a given resonant mode satisfying the above two requirements. In the second, we view the linac as a hollow cylindrical waveguide having an array of discs to slow down the waves propagating in it to match the velocity of the particles' bunch and where the electric field component of the propagating mode have the right orientation.

The two most common linac configurations used for medical and industrial applications are the *standing-wave* (SW) linac and the *traveling-wave* (TW) linac. I will use the coupled-cavity approach to describe the operation of the SW linac and keep the waveguide description for the TW linac. Thus the reader can be exposed to both approaches without the risk of repeating the analysis.

2.3.1 Standing-Wave Linacs

We have learned from the above coupled resonator analysis that a chain of N-coupled cavities will have N-modes. Each mode is characterized by a specific frequency and a field pattern with a corresponding phase shift from one cavity to next. A standing-wave (SW) linac is formed of a chain of coupled cavities where the microwave is fed into one of the cavities (normally close to the middle); see a basic schematic in Figure 2.14. The electromagnetic waves propagate from one cavity to the next leaving behind some energy as they pass by. The waves get bounced back at the two conducting walls at each end of the linac. They continue traveling back and forth and building up the fields in the cavities

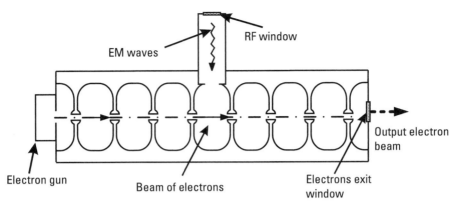

Figure 2.14 A basic schematic for an SW linac.

until the fields reach their full values. As the waves propagate in both directions and get reflected back and forth, they build up *standing-wave* patterns in the structure and, hence, the reason for the name used for this type of linac.

Let us now consider an SW linac of, for example, nine coupled cavities (such as the one we sketched in Figure 2.14). We will then have nine possible modes at which to operate. By choosing the excitation frequency that matches one of these nine modes, we can decide on which mode to operate the structure at in order to have a particle accelerator.

One possible mode for accelerating a charged particle in an SW linac would be the π mode. The corresponding field pattern for the π mode in an SW linac with nine coupled cavities can be presented schematically as shown in Figure 2.15.

In order to achieve acceleration effectively, we need the particle (or bunch of particles) to be at the center of the cavity when the field in the cavity is at its peak, and by the time the RF field has changed by 180 degrees, the particle (or the bunch of particles) has traveled to the center of the next cavity. Thus the particle (or the bunch) would always see an accelerating field with the right polarity and at the peak. The condition for this synchronism would be that the cavity length d and the particle velocity v would be related to the RF wavelength by the relation:

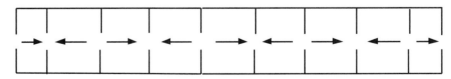

Figure 2.15 A schematic representation of π mode in an SW linac with nine coupled cavities.

$$d = \frac{v}{c}\frac{\lambda_{RF}}{2} \qquad (2.12)$$

To achieve synchronism between the particles accelerated and the RF field in a linac, it is essential that we group the particles into bunches. The phase of each bunch can then be adjusted to be at the center of the cavity when the RF filed is at its peak.

The synchronism condition in (2.12) requires that the cavity length d would be increasing from one cavity to the next to match the increase in the particles' velocity v. However, in many practical electron linacs, the electron velocity would approach the speed of light after the first few cavities. The rest of the cavities downstream would be of fixed length corresponding to the half-wavelength of the RF.

Since protons are about 2,000 times heavier than electrons, they would take much more energy, compared to electrons, before approaching the speed of light. Thus in practical proton accelerators, successive cavities have to be of increasing length, since the particle velocity v is increasing along the accelerator.

The π mode is implemented in many practical accelerators, especially those linacs made up of few cavities. However, if we keep increasing the number of cavities in our SW linac, we soon run into a problem if we are using the π mode. As we increase the number of coupled cavities in the linac, the number of modes will correspondingly increase. The modes become closer to each other in frequency. The dispersion curve shown in Figure 2.16 is for a linac with 50 coupled cavities. The frequency spacing between the π mode and the closest mode is small because the π mode and the next nearest-neighbor mode are at a region on the dispersion curve where the slope of the curve is close to zero. Thus the RF power from the generator can excite both modes rather than having all power in the desired accelerating mode, π mode, compromising the performance of the linac.

In the above dispersion curve (Figure 2.16), we notice that the $\pi/2$ mode lies in the center of the linear region and enjoys the largest mode spacing with its nearest neighboring modes. For this reason, that mode is employed in many of the SW linacs. The field pattern for an SW linac operating at this mode is shown in Figure 2.17(a). Although the linac would operate stably in this mode, this operation would not be an efficient one since every second cavity does not carry any electric field and thus produces no energy gain for the particles being accelerated. The logical solution would be to make these zero field cavities shorter, as shown in Figure 2.17(b). This configuration represents a class of SW linacs that is called *on-axis biperiodic* standing-wave linacs. It can be viewed as comprised of two groups of cavities. The normal length cavities are the *accelerating cavities,* also called the *main cavities*. The shorter cavities make up the second group. They have no field; hence they do not participate in the

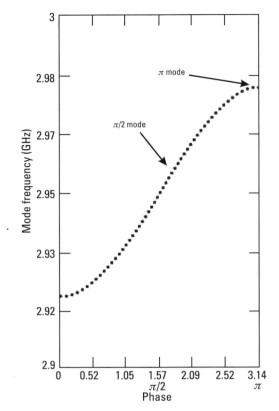

Figure 2.16 Dispersion curve for an SW linac with 50 coupled RF cavities.

acceleration. However, they participate in passing the RF power along the linac and they couple the accelerating cavities together. They are called the *coupling cavities*, or the *auxiliary cavities*. We can even make this linac shorter in length by moving the coupling cavities to the sides and away from the beam channel, as shown in Figure 2.17(c). This is the configuration of the *side-coupled* standing-wave linac [5, 6].

The fact that the coupling between the accelerating cavities is not just through slots but rather through cavities [the coupling cavities, see Figure 2.17(b, c)] enhances the coupling and makes the linac more stable. The coupling is largest when the coupling cavity is at resonance. Conversely, at other off-resonance frequencies the coupling is orders of magnitude smaller. This mechanism of coupling through the auxiliary cavities is called *resonant coupling*. Actually, we can make the side cavities smaller in size and still maintain the same resonance frequencies by adding posts in the coupling cavities as shown in Figure 2.18. The posts add to the capacitance in the coupling cavity, thus compensating for the reduction in size and maintaining the same resonant frequency.

Figure 2.18 shows a partially sectioned "side-coupled" SW linac. The cavities on the beam axis are the accelerating cavities. The cavities on the side, the coupling cavities, nominally have no field but help stabilize the linac operation against perturbations from manufacturing errors.

As we discussed above, the electrons in a linac get accelerated in bunches. We will discuss how the electrons cluster into bunches later in this chapter. For now, let us follow a bunch of electrons as the bunch gets accelerated down a side-coupled SW linac. Referring to Figure 2.19, if the electron bunch is at the center of cavity A, the bunch gets accelerated under the force of the electric field wave E, as it is in a negative excursion. The reader would recall that electrons are accelerated in opposite direction of an electric field. During that time, the E wave of the adjacent cavity B is positive and electrons would actually be decelerated if they were in cavity B at this instant. However, as the electron bunch progresses from cavity A to cavity B, the E wave has switched polarity and now

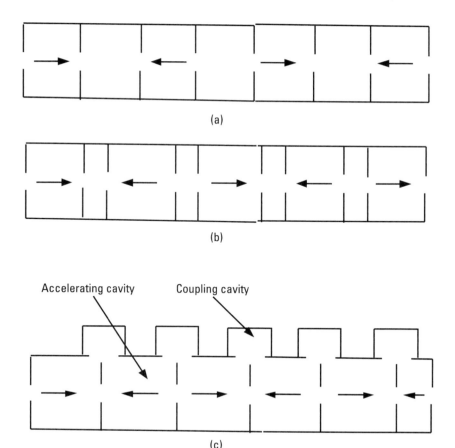

Figure 2.17 Different configurations for an SW accelerator operating in the $\pi/2$ mode.

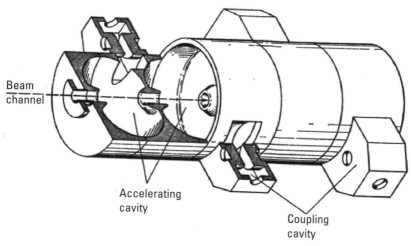

Figure 2.18 Side-coupled linac [3].

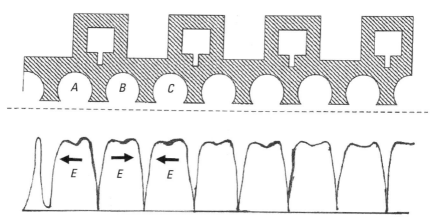

Figure 2.19 Acceleration mechanism in a side-coupled SW linac.

is in its negative excursion, and the electron bunch continues to be accelerated. This process continues until the electrons attain their final energy. Here we see the need for accurate machining and tuning of each cavity, such that the bunch is at the center of the cavity when the RF field is at its peak and with right polarity as the bunch progression from one cavity to the next.

2.3.2 TW Linacs

In the TW linac, the microwaves enter the accelerating structure on the electron-gun side and propagate toward the high-energy end of the accelerator, where they are either absorbed without any reflection or exit the accelerator to be absorbed in a resistive load, as shown in Figure 2.20. As mentioned before, we need to slow down the waves propagating in a linac to match the velocity of the particles being accelerated. This can be done in a variety of ways, but the most common approach is to insert irises or disks into the linac at a specific periodicity chosen in such a way that the accelerated particles move in step with the wave [7]. This configuration results in what is known as the *disk-loaded* traveling-wave linac. The principle of operation of the TW linac is straightforward. The electromagnetic wave is launched at the first cell of the linac; the wave propagates along the beam axis, and a beam of particles is injected from an electron gun along the axis. The electrons are grouped into bunches, which are accelerated by the force of the electric field of the electromagnetic wave. The conducting walls of the linac and the beam absorb the RF power. Thus, the RF field amplitude gets attenuated along the linac. At the end of the linac, the remaining RF power is absorbed by resistive material fused to the wall of the last cavity or delivered to an external load, as shown in Figure 2.20. In either case, no RF power should be reflected to avoid any possibility of establishing a standing wave in the linac. An analogy that is often used to convey the concept of the operation of a TW linac is that of a person surfing on an ocean wave. The surfer rides the forward edge of the crest and travels in step with wave. Similarly, the accelerated bunch of charged particles travels on the front of the advancing electromagnetic wave.

The TW linac is mostly used for accelerating electrons that would reach speeds close to the speed of light once they gain energies higher than 1 MeV, which can be attained by the third or fourth cell in many practical TW linacs. This is not the case for heavier particles, such as protons, since they travel at much slower velocities that would not allow synchronism with the electromagnetic wave in the TW linac. Another difficulty with accelerating protons or

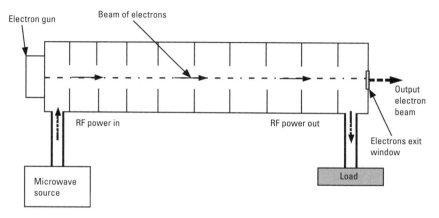

Figure 2.20 A basic schematic for a traveling-wave linac.

ions in a TW linac is that the velocity of the heavy particle would be continuously increasing as they get accelerated, which would require the TW cells to be changing in length from one cell to the other, making it harder to design and manufacture. In contrast, in many commercial TW electron linacs only the first few calls need to be gradually increasing in length and the rest of the cells would be of fixed length.

A note about the terminology related to the linac. Users of the linac in the field widely employ such jargon as *waveguide accelerator* or, even more commonly, *waveguide*. Clearly the term waveguide is a misnomer. Strictly speaking, a waveguide is a metallic pipelike section that is used in transmitting microwave power. However, it is easy to understand the origin of this lingo that has been traditionally used in the trade. As we mentioned previously, the TW linac can be viewed as a hollow cylindrical waveguide having an array of discs to slow down the waves propagating in it to match the velocity of the particle.

2.3.3 Bunching Mechanism

As we pointed out earlier, in order to be able to achieve synchronism between the particles accelerated and the RF field in a linac, we group the particles into bunches (see Figure 2.21). The phase of each bunch can then be adjusted to be at the center of the cavity when the RF field is at its peak.

To explain the bunching mechanism we use Figure 2.22, and to simplify the explanation, we will consider positively charged particles in a cavity-hosting electric field. Let us consider three particles: particle S is the synchronous particle (the center of the bunch), particle L is a late particle, and particle E is an early particle. The late particle L sees larger electric field and hence gets more boost than S. Similarly, the early particle E sees less field and gets less push. As a result they all eventually bunch together.

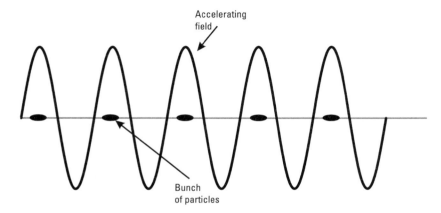

Figure 2.21 The grouping of charged particles into bunches.

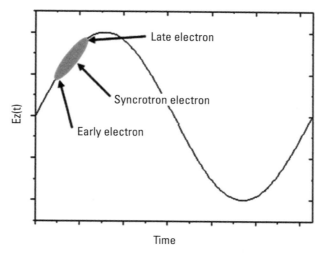

Figure 2.22 Bunching mechanism.

This process would be carried out in the first two or three cavities. Once the bunch is formed, its phase is adjusted to be "riding" on the crest of the electric field wave to get the maximum acceleration form each cavity. The same principle applies to electrons, except that the electron would be bunched and accelerated by a negative electric field. Actually, when the electrons are injected from an electron gun into the first cavity of the linac, those that enter the linac in the retarding phase (electric field positive) are rejected and would not be part of the bunches being formed.

In many medical and industrial linacs, electrons injected from the electron gun would have only energy of tens of kiloelectron volts (10–50 KeV). The electrons gain energy as they pass through the first few cavities of the linac and gradually approach the speed of light. For example, in some of the short linacs electrons gain such energy by the third cavity. In this example the accelerator designer will make the first three cavities of different lengths and shorter than the rest of the linac. Each of the three cavities will be longer than the preceding one. The fourth cavity and all the cavities following would be the standard length to achieve synchronism with a relativistic electron bunch having a speed close to the speed of light.

2.4 Electron Gun Operation

The source of electrons for any electron linac is called the *electron gun* or *E-gun* for short. The majority of E-guns are based on thermionic emission, where heat is used to free electrons from emitting surfaces. Thermionic E-guns have been used extensively for many years in diverse applications such as linacs, cathode ray tubes (CRT), microwave sources such as klystrons, and other vacuum devices.

A schematic for a typical thermionic triode electron gun is shown in Figure 2.23 [3]. The gun is comprised of a *cathode* that emits electrons when heated by a *heater*, a *grid* to control the flow of electrons, *focusing electrodes* to provide the proper electrostatic boundary conditions to shape the electron beam, and an *anode*. In many cases, the anode is the first cavity in the linac, which is obviously kept at ground potential. The cathode is enclosed in a ceramic body and well insulated, since it is kept at a negative high voltage (-10 to -50 KV). The power to the heater determines the amount of current emitted from the cathode. The grid regulates the current to the linac. In the standby mode of the linac operation (beam-off), the grid is at a negative potential (-100 to -200 V). The negative potential on the grid repels the electrons emitted from the cathode. During the beam-on mode, the grid is pulsed with positive potential (300V–500V), allowing the passage of electrons and controlling the linac current.

For the cathode to be a good electron emitter when heated it has to have a relatively low value of a parameter called the *work function* (the minimum energy needed to remove an electron from a solid to a point immediately outside the solid surface). Over the last six decades different improvements were added to the cathode, which normally is made of tungsten, to lower the work function of its emitting surface. We will list below two of the important improvements [8]:

Impregnation. The tungsten cathode is impregnated with work function–lowering chemicals. Common cathode impregnate compositions include

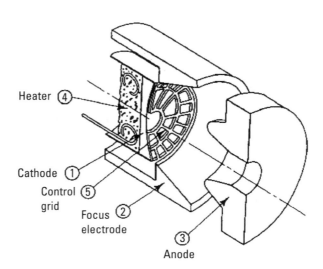

Figure 2.23 Schematic of a linac electron gun.

the mixture of barium oxide (BaO), calcium oxide (CaO), and aluminum oxide (Al_2O_3) compounds. Different mixes of these three compounds are used in E-gun cathodes, and the most commonly used have relative molar percentages of BaO, CaO, and Al_2O_3, respectively, as 5:3:2, 4:1:1, or 3:1:1. Barium atoms and barium oxide molecules are produced by chemical reactions and thermal decomposition. These gaseous species diffuse through the cathode pores to the cathode-emitting surface, lowering the emitter work function. For the barium to diffuse through the tungsten cathode, it has to be porous. Most cathodes are made of approximately 80% dense metal matrix tungsten with pore area of about 20 μm^2 and pore diameter of about 5 μm.

Emitting surface coating. An important improvement to the electron gun–impregnated cathode was employing a coating of an alloy of platinum group metals, such as osmium-ruthenium alloy (Os-Ru coating), on the emissive surface to lower the work function by approximately 20%. This type of cathode is known as *M type* cathode. As a result, this emitter can be operated at lower temperatures than the uncoated impregnated cathode for the same current density. It can operate 100°C cooler, which can result in a reduction of up to 85% in barium evaporation rate as well as a 35% saving in heater power and extending the cathode life 10 times.

An electron gun used with medical linacs is shown in Figure 2.24.

2.5 X-Ray Generation and Target Design

In many linac applications, including cancer radiation therapy, the linac output is an X-ray beam rather than an electron beam. The most common technique for the generation of an X-ray beam using a high-energy electron beam is to

Figure 2.24 Typical electron gun used with medical linacs. (Courtesy of Altair Technologies, Inc. [9].)

impinge the e-beam on a metallic *X-ray target*. The process taking place in the target in which the energy of electrons is transformed into X-ray photons is called *bremsstrahlung*, which literally means in German *breaking radiation*.

2.5.1 Mechanism of Conversion

The process of bremsstrahlung is the result of interaction between high-energy electrons and nuclei of the target material atoms. An energetic electron, while passing near a nucleus, may be deflected from its path by the action of Coulomb forces of attraction and lose energy (see Figure 2.25). As a result, a part or all of its energy is dissociated from it and propagates as electromagnetic radiation (photons). Since this interaction may result in partial or complete loss of electron energy, the resulting photon may have any energy up to the energy of the incident electron. Thus, the resulting photon beam has an upper energy limit equal to the energy of the impinging electron beam. This conversion process is not very efficient since many of the electrons do not partake in this process. The efficiency of conversion is given by the ratio of the total power of the X-ray beam to the total power of the impinging electron beam. Typically this conversion efficiency is less than 5%. Therefore we expect that a lot of heat would be generated in the target and cooling measures should be incorporated in the X-ray target design.

2.5.2 X-Ray Target Design Requirements

Most X-ray linacs use transmission-type targets in which the electrons bombard the target from one side and the X-ray beam is produced on the other side with

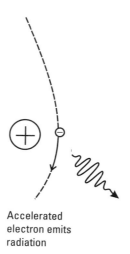

Accelerated electron emits radiation

Figure 2.25 Bremsstrahlung process of photon generation.

a relatively narrow forward cone around the direction of the incident electron beam. The higher the energy of the electron beam, the narrower the X-ray forward cone. From the above description of the conversion process, one can list some of the material requirements for an efficient target:

1. Since the bremsstrahlung is based on the deflection exerted on a passing electron by the Coulomb forces, then the larger the nucleus of the target material atoms, the more efficient the process. Hence, It is desirable to use a high atomic number (high Z material). Tungsten, gold, platinum, and tantalum are possible material choices for linac X-ray targets.
2. Since the process has a limited efficiency and many of the impinging energetic electrons do not contribute their energy to the conversion, a lot of energy is dissipated as heat into the target. For this reason the target material should have a high melting point, such as refractory metals. Tungsten and tantalum are good examples of refractory metals that are used as X-ray targets.
3. It is preferable that the target material has good thermal conduction to facilitate the transfer of the heat generated to the cooling medium incorporated in the target design.

2.5.3 Target Design Approaches

Different target design approaches have been implemented in medical and industrial linacs. The target can be internal or external to the linac. It can be a fixed or moving target. Since not all the electrons contribute to conversion process, some electrons pass through the target. Provisions are usually provided downstream from the target to absorb such electrons in order to have a non-contaminated beam of X-ray. Therefore, the target layer is usually followed by a layer of low atomic number (low Z) material, such as carbon or aluminum, in which any remaining electrons are absorbed and filtered out of the X-ray beam.

Internal Versus External Target

In some linac designs, the target is an integral part of the linac and included within its vacuum envelope. In others, the target is a separate component outside the vacuum envelope and gets attached downstream of the linac. The later approach would require the linac to have a beam *exit window* that would seal the vacuum but allow the passage of the electron beam impinging toward the external target with minimum loss to the beam. Some linacs use titanium or similar metal as the exit window. The thickness should be a tradeoff between durability (thicker window is preferred) and beam attenuation (thinner window

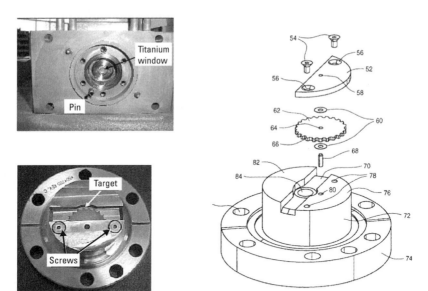

Figure 2.26 Drawing for a spinning target by Siemens [10].

is preferred). For many linacs, an exit window thickness of 50–150 μm would be a good trade-off.

Fixed Versus Moving Target

It is a simpler design to have the target made of nonmoving parts. In these configurations, a high thermal conductivity material such as copper is incorporated in the target design to help draw off heat generated in the high Z target layer. However, this approach is not sufficient in some linac applications with high beam intensity. In a fixed target the heat generated is concentrated in an area equal to the beam spot size at the target (a couple of millimeters in diameter in many cases). This can cause concentrated thermal stresses as well as mechanical fatigue in the target layer. An alternative approach is to distribute the location of beam incidence on the target by having the target spinning around an axis off the beam line. Thus the locus of beam incidence would be in the form of ring of width equal the beam diameter at the target. Figure 2.26 is extracted from one of the U.S. patents by Siemens Medical Systems, Inc. [10]

References

[1] Lapostolle, P. M., and A.L. Septier (eds.), *Linear Accelerators,* North Holland Publishing Co., Amsterdam, the Netherlands, 1970.

[2] Wangler, T. P., *Principles of RF Linear Accelerators,* New York: John Wiley and Sons Inc., 1998.

[3] Karzmak, C. J., et al., *Medical Electron Accelerators,* New York: McGraw-Hill Inc., 1993.

[4] Pozer, D. A., *Microwave Engineering,* Addison-Wesley Publishing Co., 1990.

[5] Nagle, D. E., E. A. Knapp, and B. C. Knapp, "Coupled Resonator Model for Standing Wave Accelerator Tanks," *Rev. Sci. Instr.,* Vol. 38, 1967, pp. 1583–1587.

[6] Knapp, E. A., B. C. Knapp, and J. M. Potter, "Standing Wave High Energy Linear Accelerator Structures," *Rev. Sci. Instr.,* Vol. 39, 1968, pp. 979–991.

[7] Loew, G. A., and R. B. Neal, "Accelerating Structures," *Linear Accelerators,* P. M. Lapostolle and A. L. Septier (eds.), Amsterdam, the Netherlands: North Holland Publishing Co., 1970, pp. 39–113.

[8] Cronin, J. L., "Modern Disperser Cathodes," *IEEE Proc.,* Vol. 128, 1981, pp. 19–32.

[9] Altair Technologies Inc., Menlo Park, CA.

[10] Siemens Medical Systems Inc.. 1998. Rotary target driven by cooling fluid flow for medical linac and intense beam linac. US Patent 5,757,885, May 26, 1998.

3

Linac Supporting System

In Chapter 2, we discussed the fundamental concepts of operation of different types of the linear accelerator (or linac) and its main constituents. In this chapter we describe other essential auxiliary components of the linac and a typical linac-based RF system such as those used in machines for medical and industrial applications. Some of these applications are discussed in Chapter 5 and Section 6.3.1.

3.1 Introduction—The Linac as a Source for Electron and X-Ray Beams

Medical and industrial linac applications use machines built around a linac that generates high-energy electrons beams or high-energy X-rays needed for cancer treatment or industrial processing. In Figure 3.1, we show a basic presentation of the linac as a source for X-rays [1].

As we discussed in Chapter 2, the main constituents of an electron linac as a source of X-rays are:

1. The electron gun, where the electrons are generated;
2. The main linac body, which is comprised of a series of coupled cavities where the electrons are bunched and accelerated;
3. The X-ray conversion target, where the X-rays are generated.

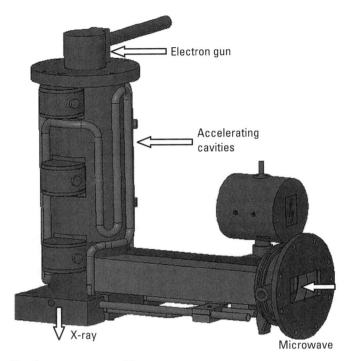

Figure 3.1 The linac as a source of X-rays.

The operation of the linac requires an RF source to provide the microwaves whose fields accelerate the electrons, the means for achiving a good vacuum in the linac, and a system for water cooling to maintain the linac temperature for a stable operation of the linac. In Figure 3.2, we show schematically the main consiuents of a linac (E-gun, accelerator body, and the X-ray target).

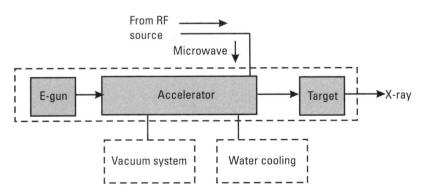

Figure 3.2 Linac main constituents and auxiliary systems.

Additionally, we show two of the auxiliary subsystems, namely, the linac's vacuum system and the water-cooling system.

3.2 Linac Auxiliary Systems

3.2.1 Linac Vacuum System

There are three reasons to keep the linac under high vacuum. First, the particles being accelerated (electrons, protons, or ions) should be able to travel free of any deflections caused by collision with gas atoms or molecules in the linac. Second, the high electric field needed to accelerate the particle beam can ionize any remnant atoms or molecules floating inside the linac proper, resulting in arcing and breakdown. Third, it is important to avoid the precipitation of impurity gases on the surface of the electron gun's cathode. This may cause contamination of the cathode, known as *cathode poisoning*. Such precipitation of foreign atoms and molecules on the cathode surface causes electrochemical reactions that can reduce the efficiency of emission from the cathode surface and, ultimately, reduce the useful life of the electron gun.

It is therefore desirable to maintain in a linac a vacuum level better than 10^{-6} Torr. In fact, many linac manufacturers build accelerators with vacuum better than 10^{-8} torr. The torr is a unit for describing gas pressure. It corresponds to 1/760 of the standard atmospheric pressure. Since it is a relatively small unit of pressure, it is commonly used to describe vacuum levels.

To achieve good vacuum in a linac, four points have to be fulfilled. They are:

1. The linac should be well sealed to ensure no *external leaks*;
2. The inner surfaces of the linac should have low *outgassing* (release of trapped gas atoms and molecules) rates;
3. The linac inner surfaces should not have pockets of trapped gas, which cause what is known as *virtual leaks*;
4. The linac should be connected to a *vacuum pump* with a sufficient pumping speed;
5. The linac design should provide for adequate *vacuum conductance* between different sections of the linac.

In medical and industrial accelerators an *ion pump* is normally connected to the waveguide section between the RF window and the cavity in the linac to which the RF is fed (see Figure 3.3). The ion pump serves to maintain high vacuum through gas ionization. It is connected to a high-voltage power supply (3–5 KV) to provide the voltage difference necessary for the ionization process

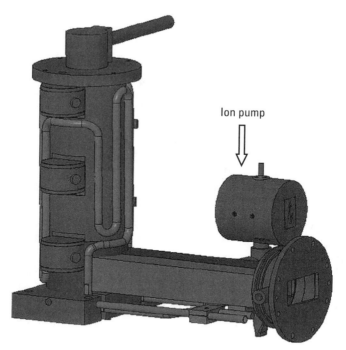

Figure 3.3 Linac with an ion pump for maintaining its vacuum level.

to take place inside the pump [2]. By monitoring the current drawn from the power supply, we can gauge the vacuum pressure of the system. In many linac applications, the ion pump current is used as a parameter to interlock the radiation system. So in case of vacuum failure due to excessive outgassing, vacuum leak, or ion pump malfunction, vacuum control interlock will stop the linac irradiation.

3.2.2 RF Vacuum Window

In the basic linac shown in Figure 3.4, a component is marked as RF window at the port where the RF is fed into the linac. The function of the RF window is to separate the linac vacuum on one side from the pressurized gas (such as sulfur hexafluoride, SF_6) in the transmitting waveguide on the other side. At the same time, the RF widow provides a good RF match for electromagnetic transmission. The reason for filling the waveguides with pressurized gas is to prevent arcing in the waveguides when they are transmitting high-power microwave from the RF source to the linac, as will be discussed in Section 3.3.2 on RF Power Transmission Subsystems.

The RF window is made of ceramic, which is normally a high-purity aluminum oxide (Al_2O_3), also known as alumina, which is characterized by its

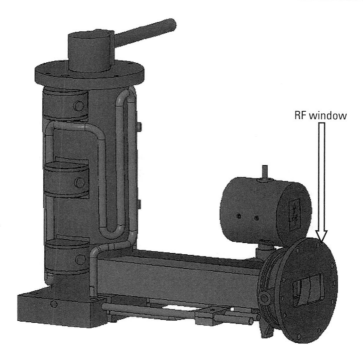

Figure 3.4 The vacuum RF window attached to a linac.

low RF loss and low outgassing in vacuum. In many RF windows, the ceramic is coated with a thin layer of titanium nitride (TiN) on the vacuum side to reduce secondary electron emission from the inner surface of the ceramic. An RF window for an S-band linac is shown in Figure 3.5 [3].

Figure 3.5 An RF window for an S-band linac. (Courtesy of Altair Technologies, Inc. [3].)

Failure of the linac's RF window is rare. However, if the window fails it could seriously damage the linac, since this can result in loss of vacuum integrity and leak of the pressurized gas from the transmitting waveguide into the linac. The most common failure mode is thermal-stress failure resulting from excessive localized heat. The sources of window heating are:

1. *Dielectric loss.* The ceramics used in RF windows such as aluminum oxide are good insulators so normal *conduction loss* (representing the flow of actual charges through the dielectric) is very low. However, all dielectrics (except vacuum) have another mechanism for loss, which is the *dielectric loss.* This is caused by movement or rotation of the atoms or molecules in an alternating electric field in the dielectric. The reader may realize that dielectric loss in water is the mechanism we exploit for heating food and drink in a microwave oven.

2. *Resistive loss.* This loss mechanism can take place in the titanium nitride (TiN) coating layer. For this reason, it is important to control the thickness of the TiN film deposited on the ceramic of the RF window. This loss mechanism can also exist in any conducting impurities embedded in the RF window ceramic. In view of that, the purity of the dielectric ceramic in the RF window is an essential requirement especially for high-power linacs.

3.2.3 Linac Water-Cooling System

Stable operation of a linac requires maintaining the frequency of operation at its nominal value. Actually, in copper linacs the power dissipated as heat in its structure is generally comparable to or greater than the beam power in the linac [4]. This heat, through thermal expansion, causes the linac dimensions to change. As a result, the linac's operating frequency drifts away from its nominal value. For example, the frequency of a linac operating in the S-band (2–4 GHz) would decrease about 50 KHz for a temperature rise of one degree C. For linacs operating in the X-band frequency range such as those operating at 9 GHz, the decrease in linac's frequency would be 150 KHz for a temperature rise of one degree C. The sensitivity of the linac's frequency to temperature varies depending upon its design and the frequency band within which the linac operates.

To maintain a precise linac temperature control, a cooling water system is needed. In Figure 3.6, we show the linac with its water cooling channels. Cooling water is supplied by a specially controlled water-cooling system. The water-cooling system should be free of scaling, corrosion, or bacteria growth. For this reason, it is common to use low conductivity deionized (DI) water.

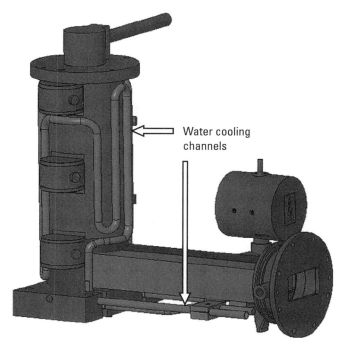

Figure 3.6 Linac water-cooling channels.

The water-cooling passages are occasionally designed to be integrated within the linac's copper structure. However, the majority of linac designs use water pipes that are brazed to the outer surface of the linac.

One of the accepted choices for water temperature in a linac cooling system is 40°C. Operating the linac at a temperature that is above the ambient temperature makes the operation of the linac less sensitive to ambient temperature fluctuations. Additionally, having the cooling water's temperature considerably above the heat exchanger's water temperature would mean more efficient heat transfer.

3.3 Radio Frequency (RF) System

Linac-based machines for radiation therapy or industrial applications utilize the linac as the source of electron or X-ray beams. In such machines, the linac is part of an RF system similar to the one shown in Figure 3.7. The main function of the RF system is to *generate* and *deliver* to the linac the microwave power needed to accelerate its electron beam. In addition to the microwave source

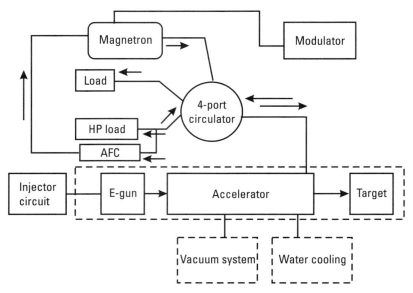

Figure 3.7 An example of an RF system for an X-ray radiation unit.

(klystron or magnetron), the RF system is comprised of the waveguides needed to transmit the RF power, a circulator that controls the flow of the RF power and protects the source from any reflected power, high-power load to absorb the reflected power, a modulator to provide the high current and high voltage pulses to the microwave source, and an automatic frequency control (AFC), which is basically a feedback system to synchronize the frequency of the source with that of the linac.

3.3.1 RF High-Power Sources

There are two types of high-power RF sources that are commonly used to provide the microwave to medical and industrial linacs. One source functions as an "oscillator," which is the *magnetron*, and the other is a high-power amplifier, the *klystron*.

3.3.1.1 The Magnetron

The magnetron is used in many of the radiation therapy or industrial units, especially with the units employing linacs with electron-beam energy less than 10 MeV. In Figure 3.8, we show as an example an S-band magnetron MG5193, a product of e2v [6]. Peak output power of this pulsed magnetron is 2.6 MW, and it is mechanically tuned over frequency range 2,993 to 3,002 MHz.

The mechanism of any high-power RF source is based upon two fundamental requisites. These are the existence of accelerated electrons (in vacuum)

Figure 3.8 MG5193 tunable S-band magnetron by e2V [6].

and resonant cavities. DC voltages and currents are required to provide the energy needed to accelerate the electrons. If the source is an oscillator such as the magnetron, no input RF signal is needed in general. This simplified concept is depicted in Figure 3.9.

The basic structure of the magnetron is an *anode* made of a number of identical cavities arranged in a cylindrical pattern around a central cylindrical *cathode*, as shown in Figure 3.10, and the space between the cathode and anode is evacuated [7, 8]. The cathode is indirectly heated by an inner tungsten filament and the electrons are generated by thermionic emission, the same emission mechanism discussed before in our description of the linac's electron gun. When the magnetron is pulsed, the anode is at a high positive voltage V_a relative to the cathode. Also, the whole structure is immersed in a static magnetic field provided by two poles of an electromagnet. Thus, a uniform magnetic field B_0 is applied parallel to the axis of the cylindrical cathode and anode. The electrons emitted from the cathode are then accelerated toward the anode under

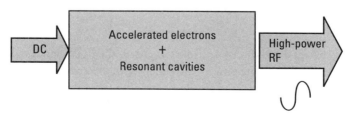

Figure 3.9 Simplified fundamental concept of the magnetron.

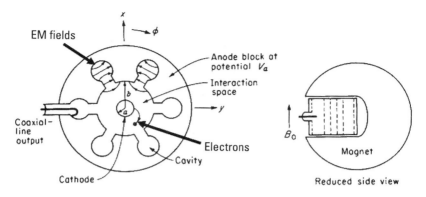

Figure 3.10 Electric field and magnetic field in a magnetron [7].

the forces of a radial electric field, but the presence of the magnetic field B_0 produces an azimuthal force in the ϕ direction. The combined electrical and magnetic forces make the electron to follow a curved trajectory, as shown in Figure 3.10. As the electrons move close to the anode cavities, they radiate some of their energy in the form of microwaves. The cavity geometry and dimensions determine the central frequency of the magnetron. The generated RF pulses are coupled out of the anode cavities via a loop to the external waveguide through a glass window. The cylindrical waveguide then connects to the rest of the RF system of the radiation unit.

Since anode cavities interact together as coupled resonators, they can oscillate at different modes. The usual mode employed in a magnetron is the π mode, where the phase difference between adjacent cavities is 180°. A sketch of the electric field lines in two adjacent cavities is shown in Figure 3.10. To reduce the possibility of oscillation in other possible modes, magnetron manufacturers use *strapping*. The straps connect alternate cavities that are of equal phase and pass over adjacent cavities which, at mode frequency, are 180° out of phase, to ensure the operation of the magnetron in the π mode. In Figure 3.11, we show a cutaway magnetron picture and a schematic diagram to illustrate different parts of the magnetron. The figures show the strapping as well as the *output loop* used to couple the output RF power from the anode cavities to the cylindrical waveguide acting as the output port of the magnetron.

3.3.1.2 The Klystron

The klystron is used as the RF source in many of the linac-based radiation units, especially with these units employing linacs with electron-beam energies larger than 10 MeV. It is used as an RF power amplifier that amplifies its input of low-power RF, which is generated by a low-power RF source commonly called the *RF driver*. Similar to other high-power microwave tubes, and as we pointed out

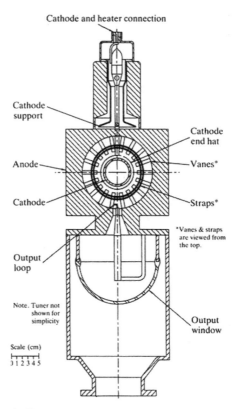

Figure 3.11 A schematic diagram to illustrate different parts of the magnetron.

for the magnetron, we need the existence of accelerated electrons (in vacuum) and resonant cavities. DC voltages and currents are needed to provide the energy required for accelerating electrons. This basic concept behind the klystron is depicted in Figure 3.12.

The simplest klystron has two resonant cavities: the first cavity, called the *buncher cavity*, which is energized by low power microwaves from the RF driver, and the second cavity, called the *catcher cavity*, which extracts the microwave power from an energetic beam and emits the high-power RF output. A cross-sectional drawing of an elementary two-cavity klystron is show in Figure 3.13. The cathode receives dc pulses from a special power supply called the *modulator*. These pulses make the cathode at a negative potential with respect to the cavities. Electrons produced by the heated cathode would then accelerate toward the buncher cavity and arrive with uniform velocity. When electrons pass through the gap in the buncher cavity, they are either accelerated or decelerated by the oscillating RF field generated in this cavity, which is fed from the RF driver. As a result, some of the electrons would get faster while others

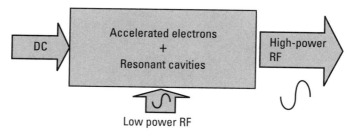

Figure 3.12 Simplified fundamental concept of the klystron.

would be slowed down. Thus, the velocity of the electrons is altered by this action, resulting in *velocity modulation* of the electron beam to gradually bunch together as they travel in the drift space between the buncher and the catcher cavities. When these electron-bunches arrive at the second cavity (catcher cavity), they induce charges on the ends of the cavity and thereby generate a retarding electric field. Under the influence of this field, the electrons decelerate and some of their kinetic energy is converted to the electromagnetic energy in the catcher cavity. The microwave power generated in the catcher cavity is coupled out to the rest of the RF system powering the linac. The electrons then continue moving past the catcher cavity with reduced velocity and deposit the rest of the kinetic energy into the *collector*, where it is converted to heat [8, 9].

One example of the klystrons used in radiation therapy is the Thales' S-band klystron TH 2157, shown in Figure 3.14. It produces pulsed microwave power with a central operating frequency of 2998.5 MHz, output peak power of 7.5 MW, pulse width of 6 μS max, and average power of 8 kW. Typically, It operates with beam current of 105A and cathode voltage of 150 kV, which are furnished by the modulator in the RF system.

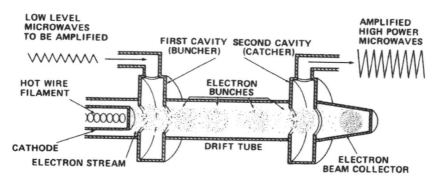

Figure 3.13 Conceptual schematic for the operation of the klystron [9].

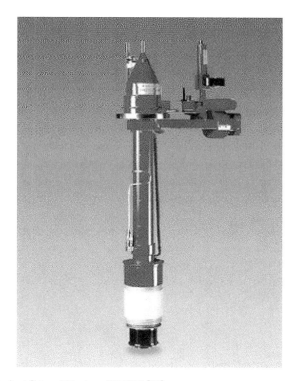

Figure 3.14 Thales' S-band klystron TH 2157 [10].

A note about the choice of magnetron versus klystron for linac-based radiation units:

Most of the radiation units using low- to mid-energy linacs (4–10 MeV) use magnetrons, which are capable of furnishing the peak power needed (less than 5 MW). Compared with klystrons, the magnetrons are physically smaller and run at lower cathode voltages, so they do not need extra electrical insulation, such as an oil tank. They also do not require an RF driver. Therefore, magnetrons can be mounted in moving radiation units such as the rotating gantries in radiation therapy machines, thus simplifying their designs. An additional advantage of magnetrons over klystrons is their lower prices.

On the other hand, klystrons are generally used in high-energy linacs (10–25 MeV) where power over 5 MW peak is needed, and they are generally more stable than magnetrons. Klystrons are larger in size than magnetrons, operate at higher voltages, require an RF driver input signal, and their cathode must be mounted within a tank of insulated oil.

3.3.2 RF Power Transmission Subsystems

3.3.2.1 Circulator

The circulator is a passive microwave component with three or four ports [8, 11], where power is allowed to flow in one direction and not in the reverse, as shown symbolically in Figure 3.15. Thus, microwave power applied to one port will only emerge at the next port.

The power flow for the three-port circulator is illustrated in Figure 3.16(a): power entering port 1 exits at port 2, while port 3 is isolated (decoupled) and sees no power out; power entering port 2 exits at port 3, while port 1 is isolated; and power entering port 3 exits at port 1, while port 2 is isolated. The same pattern applies to the four-port circulator as shown in Figure 3.16(b).

If we go back to Figure 3.7 for the overall RF system, we see that the circulator is placed between the RF source, the magnetron (or klystron), and the linac. If it is a four-port circulator, the power flow is consistent with the notation marketing in Figure 3.16. The four ports of the circulator are connected to the source at port one, the linac at port two, a high-power (H-P) load to port three, and a low-power matched load to port four. In this configuration

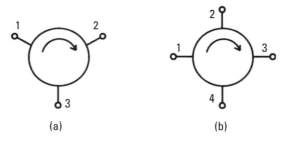

Figure 3.15 Symbolic representation of both (a) three-port circulator and (b) four-port circulator.

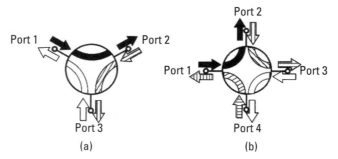

Figure 3.16 Power flow in (a) a three-port circulator and in (b) a four-port circulator. (© Ferrite Microwave Technologies, Nashua, NH.)

the RF power flows from the magnetron (port 1) to the linac (port 2). Any power reflected from the linac would flow to the high-power load (port 3). Any power reflected from the high-power load would flow to the low-power load connected to port 4. The circulator thus prevents any reflected power from being returned to the RF source (magnetron or klystron). Thus, the source would always see a matched load independent of the accelerator operating conditions. Additionally, the circulator would also prevent any build-up of standing wave in the waveguide transmission line, which can potentially cause arcing. As an example of high-power four-port circulator used in radiation therapy machines, we show, in Figure 3.17, a circulator made by Ferrite (Nashua, NH).

3.3.2.2 Waveguide Transmission

The waveguide transmission system is used to transport the RF power from the magnetron (or klystron) to the linac in radiation units. Conventional transmission lines (wires and cables) develop too much loss at microwave frequencies. Metallic hollow waveguides present an alternative at these frequencies [8, 11]. They are made of metals with good electrical conductivity, such as copper, brass, or aluminum. The waveguides used in a linac-based radiation machine would commonly have a rectangular or circular cross-section as shown in Figure 3.18. In this figure the solid arrows represent the ordination of the electric fields propagating down the waveguide. The dashed arrows represent the ordination of the electric fields a half-wave length away, 180° out of phase.

Depending on the operating frequency, the common types of waveguides used in radiation units are:

Figure 3.17 A high-power four-port circulator by Ferrite (Nashua, NH).

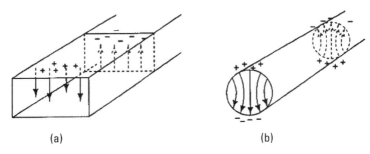

Figure 3.18 (a) Rectangular and (b) circular waveguides.

- The rectangular waveguide for S-Band (2–4 GHz) is "WR 284," where the number 284 represents the inner wider dimension (in hundredths of an inch);
- For the C-Band (4–8 GHz), the waveguides "WR 187" and " WR 137" are used;
- For the X-Band (8–12 GHz), the waveguides "WR 112" and "WR 90" are used.

In many RF systems of radiation machines, waveguides change directions at several points. For this reason, we need to include in the RF transmission system special waveguide components known as waveguide *bends,* such as those shown in Figure 3.19. The concept behind these bends is to keep the cross section uniform to avoid reflections, which usually result from a change in the waveguide cross section.

To accommodate slight misalignments between components in linac RF systems, to reduce stress on the components, and to allow for easier manufacturability, *flexible waveguide* sections are usually incorporated in the RF power transmission system. Figure 3.20 shows a cut-through view of a corrugated flexible waveguide.

The RF transmission system (including the circulator) is filled with pressurized gas, such as the sulfur hexafluoride (SF_6). Its main purpose is to prevent RF breakdown (arcing) in the waveguide transmission system. A by-product benefit is that the gas helps cool the system. The SF_6 sections are separated from the vacuum in the magnetron by its glass dome and from the vacuum in the linac by a ceramic window. Similarly, a vacuum window is used when the RF source is a klystron. The operating pressure is typically twice the atmospheric pressure. The max RF power that can be transmitted through a pressurized RF system is enhanced by a factor of 4 when using SF6 versus air at the same pressure.

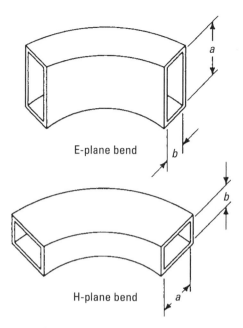

Figure 3.19 Common waveguide bends.

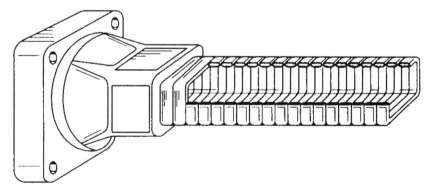

Figure 3.20 A cut-through view of a corrugated flexible waveguide.

As a protective provision, the pressurized-gas system normally includes a pressure gauge that causes an interlock, inhibiting the generation of the high-power RF if the gas pressure drops below a specified value.

References

[1] Accuray Inc., Sunnyvale, CA, http://www.accuray.com/.

[2] Varian Inc. "Vacuum Technologies," http://www.ptbsales.com/manuals/varian/ion-pumps.pdf.

[3] Altair Technologies Inc., www.altairusa.com/.

[4] Wangler, T. P., *Principles of RF Linear Accelerators*, New York: John Wiley and Sons Inc., 1998.

[5] Siemens Medical, http://www.medical.siemens.com/.

[6] Data Sheet for the Magnetron MG5193 produced by e2V www.e2v.com/.

[7] Collin, R. E., *Foundations for Microwave Engineering*, 2nd. Ed., McGraw-Hill Inc., 1992.

[8] Liao, S. Y., *Microwave Devices and Circuits*, Englewood Cliffs, NJ: Prentice-Hall Inc., 1980.

[9] Karzmark, C. J., and R. J. Morton, "A Primer on Theory and Operation of Linear Accelerators in Radiation Therapy," U.S. Department of Health and Human Services, Bureau of Radiological Health, 1981.

[10] Data Sheet for the Klystron TH 2157 produced by Thales, www.thalesgroup.com/.

[11] Pozer, D. A., *Microwave Engineering*, Addison-Wesley Publishing Company, 1990.

4

Manufacturing Techniques of Accelerators

4.1 Overview of Manufacturing Processes

Precise operations of medical and industrial linear accelerators impose certain demands on both materials used and fabrication techniques utilized. They both need to be compatible with ultra-high vacuum, high-power RF, and the presence of particle beams. In this chapter, I will briefly discuss materials compatible with these demands and then I will lead you in a guided tour through the sequence of processes of manufacturing a typical commercial linac. After selecting the right material that would constitute the cavities of the linac, our manufacturing tour starts with the special machining used to cut the linac cavities. We then stop by the chemical cleaning process, since the linac has to sustain ultra-high vacuum as well as hosting very high electric fields. Once we are done with chemical cleaning, we will move with the clean parts to a clean room, where we will observe the precise assembly, joining the parts, tuning the linac, mounting the source of electrons (the electron gun), and sealing the linac. The linac then will be put in a chamber where it will be pumped down to evacuate the sealed linac to the desired vacuum level while the linac is kept at temperatures high enough to free up gases trapped on or in the inner surfaces of the linac. Before we can start supplying full RF power into our linac, we need to do two more steps. First, prepare the electron gun for emitting a steady flow of electrons; we call this step gun activation. Second, we need to condition the inner surfaces of the accelerator to be able to stand high fields without arcing or breaking down. This conditioning step is done by increasing the RF power input to the linac slowly and gradually. Now, we can safely test the linac at the full operating power it is designed for.

Although I will focus exclusively on the manufacturing of conventional room-temperature copper linacs, many of these manufacturing techniques apply also to the manufacturing of superconducting niobium linacs, except for a few special processes. Readers interested in superconducting linacs can consult some of the dedicated publications, such as the excellent references [1, 2].

As an overall guide, I have mapped for you the flow of the linac manufacturing processes in Figure 4.1. Now we should be ready to learn about each of these processes; let us begin.

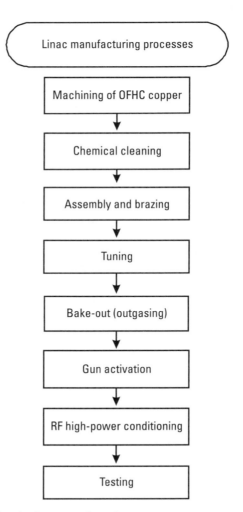

Figure 4.1 Process flow for linac manufacturing.

4.2 Material Requirements

The proper operation of a linac depends on several factors. One significant factor is the selection of the material used in fabricating the linac cavities. Some of the requirements for the choice of such material are as follows:

1. *Electrical conductivity.* This is the most important parameter for the linac material. For a given geometry and frequency, this parameter determines the power loss of the linac's walls and hence the linac's efficiency. For this reason, it is always recommended to use OFE (Oxygen-Free Electronic) copper. Another common name linac manufacturers use is OFHC (Oxygen-Free High-Conductivity) copper. The oxygen content in the OFE copper is normally restricted to 5 parts-per-million (ppm). The electrical conductivity of OFE copper is around 5.8×10^7 1/(ohm-cm). This is superior to aluminum [3.7×10^7 1/(Ohm-cm)] or stainless steel [1.5×10^6 1/(ohm-cm)] [3]. Although silver has an electrical conductivity of 6.2×10^7 1/(ohm-cm), which is better than copper, copper is used more commonly in linac manufacturing for obvious economical reasons.

2. *Thermal conductivity.* This is the second important parameter in choosing the linac material. It determines the rate of transfer of heat into the cooling system for a given power dissipation in the linac's walls. Again, copper is chosen because of its high thermal conductivity, 4.0 (W/cm/°C). The electrical and thermal conductivities for OFE copper, aluminum, stainless steel, and silver are listed in Table 4.1 [3]. One can notice from the table that the values of electrical conductivity and of thermal conductivity are closely correlated for these metals.

3. *Vacuum outgasing.* The rate of outgasing from the linac inner surfaces determines the sustainable vacuum levels in the sealed linac for a given installed vacuum pump. Thus, materials used for the construction of linacs should have low outgasing (release of trapped gas atoms and

Table 4.1
Electrical and Thermal Conductivities for Linac Materials

Material	Electrical Conductivity 1/(ohm-cm)	Thermal Conductivity (W/cm/°C)
OFE copper	5.9×10^7	4.0
Aluminum	3.7×10^7	2.2
Silver	6.2×10^7	4.1
Stainless steel	1.5×10^6	0.2

molecules) rates. Copper used for linac cavities are normally mechanically worked by rolling or extruding to eliminate porosity and reduce grain size, which help reduce outgasing rates.

4. *Mechanical stiffness and machinabilty.* This property affects the elastic deformation of the linac cavities under external pressure. Unintended deformation can perturb the linac frequency. The parameter quantifying this property is Young's modulus of elasticity. Additionally, the ease of machining the material is another consideration. Copper can be cut quickly and easily to obtain a good surface finish without wearing down the cutting tool much.

   Although OFE copper is soft after annealing, it work-hardens rapidly when distorted. For this reason, we will see when we discuss the tuning of cavities in Section 4.6 of this chapter, that tuning needs to be performed in as few operations as possible because the work-hardening requires the use of greater and greater force on each successive operation.

5. *Cost.* Obviously, this factor should be considered in deciding on which material to use for making commercial linacs. For example, we saw above that the electrical and thermal conductivities of silver are superior to copper. However, as a trade-off almost all linac manufacturers use copper, given the price difference.

6. *Electron secondary emission yield.* The electron secondary emission yield (SEY) describes the number of electrons emitted from the inner surfaces of the linac for each electron incident on the surface. Copper has a comparatively low secondary emission yield.

7. *Other requirements.* We briefly listed above the general linac material requirements. However, special linac applications may compel the linac manufacturer to consider additional material properties. Examples are: magnetic permeability, mechanical creep resistance, and mechanical fatigue.

Given the above requirements, it has been established that because of its high electrical conductivity, the OFE copper is the preferred material for room-temperature linacs. Additionally, it can be machined to high precision and good surface finish, it is compatible with ultra-high vacuum (UHV), is an excellent heat-conductor and can be conditioned with high power RF to prevent RF breakdown, and it has a comparatively low secondary emission yield. Its main disadvantages are its relatively low strength and that it becomes soft after thermal annealing at high temperatures.

4.3 Cavity Machining

The preferred machining operation for linac cavities is the lathe turning. Fortunately, most linac cavity designs have cylindrical symmetry and thus lend themselves to turning as the preferred machining operation for linac cavity cutting. Most of the machines used are of the computed numerically controlled (CNC) type. Coolant and machining lubricants used in this operation should be water-based, not oil-based. These machines should have filters to continually filter the machining fluids, which also should be changed periodically. This machining process should result in surfaces with no sharp ridges or burrs. The roughness of the resulting surfaces is characterized by a parameter called the average roughness (Ra), expressed in one of two units: the microinch (μ''), which is one-millionth of an inch, or the micrometer (μm), which is one-millionth of a meter. Basically, the Ra is the average difference between the peaks and valleys of the surface over a certain length. Typically, the resulting surface roughness is about Ra = 12 – 16 μ'' (0.3 to 0.4 μm). A well-controlled turning machining process, including the choice of tool, feed per spindle rotation and speed of rotation, machine geometry, degree of vibration isolation, and environmental conditions can result in surface roughness of Ra = 8 – 10 μ'' (0.2 to 0.25 μm). These surface finishes are normally obtained using polycrystalline diamond tool bits. For even more superior surface finish, Ra ≈ 2 – 4 μ'' (0.05 to 0.1 μm), single-crystal diamond is used as the cutting tool and sophisticated turning techniques are employed [4].

4.4 Chemical Cleaning

Since the linac inner proper will be under high vacuum, it is important that the inner surfaces be clean and that most of the gases adsorbed on or absorbed in the surfaces are removed. Actually, on every exposed copper surface, a thin oxide layer forms, and it is usually this porous oxide layer in which gases are easily trapped. For this reason, it is important that the chemical cleaning process involves removing the oxide layer from the inner surfaces of the linac cavities. One possible chemical cleaning process for OFE copper is used in European Organization for Nuclear Research (CERN) [5] and is as follows:

1. Vapor degreasing in percholroethylene (C_2Cl_4) at 121°C.
2. Alkaline soak with ultrasonic agitation for 5 minutes in an alkaline detergent at 50°C.
3. Tap water rinse.
4. Pickling in HCL (33%) by volume with H_2O 50% at room temperature for 1 to 5 minutes.

5. Tap water rinse.
6. Acid etch in:
   H_2SO_4 (96%) 42% by vol.
   HNO_3 (60%) 8% by vol.
   HCl (33%) 0.2% by vol.
   H_2O to complete to 100%
   This etch is done at room temperature for 30 seconds to 1 minute.
7. Tap water rinse.
8. Passivation in
   H_2CrO_4 (Chromic acid) 80 g/l
   H_2SO_4 (Sulfuric acid) 3 cm³/l
   This is done at room temperature for 30 seconds to 1 minute.
9. Running tap water rinse
10. Cold deionized (DI) water rinse.
11. Dry with filtered air or dry filtered N_2.
12. Wrap in Al foil.

4.5 Assembly and Bonding Techniques

After chemical cleaning, the clean cavities and other linac components are moved (wrapped in aluminum foil or lint-free tissue) from the chemical cleaning room to the clean room, where they will be assembled. Extra care should be given to this step to ensure the clean handling of parts, such as wearing the proper clean room attire and masks, and using clean gloves, as shown in Figure 4.2

The techniques commonly used to bond the cavities are brazing and diffusion bonding. In both techniques, extra attention should be paid to ensure precise alignment at assembly to produce an accurately straight linac.

4.5.1 Brazing

The prevailing technique for joining the linac's constituent cavities and parts is brazing. The brazing process involves the use of brazing fillers (for example copper-gold alloy) that are inserted into the joints between adjacent cavities before mounting them in a brazing furnace. Thus, brazing is different from welding, where two parts of the same metallic material are heated locally above the melting temperature of the metal. In brazing, the components to be joined are placed in a furnace such as that shown in Figures 4.3 and 4.4. The brazing furnace provides heat to melt the brazing filler that solidifies later to form the

Figure 4.2 Assembly of linac parts in a clean room [6].

vacuum-tight joints. The filler can be a metal or an alloy that has a melting temperature lower than the two parts to be joined. The filler can be in the form of brazing foil or a brazing wire. During the brazing cycle the joint is heated to a certain temperature lower than the melting point of the parts but high enough to melt the filler material. The brazing filler then melts and wets the joining surfaces flowing by the capillary action through specifically designed gaps and small channels. In the cooling phase of the brazing cycle the filler solidifies, creating a strong bond at the joint. It is advisable to perform leak tests of subassemblies once they are brazed before getting to the next assembly step. One of the common vacuum-leak checking techniques is to use helium leak detectors [7]. With this technique the operator can spot the points of leak, and the part can be rebrazed and rechecked for leak before proceeding to the next assembly step.

The conditions for a good braze are as follows:

- The melted filler material wets the joint surface well and in a controlled way.
- The pieces to be joined do not "float" out of their position while the brazing alloy is the liquid phase.

Figure 4.3 A linac brazing furnace.

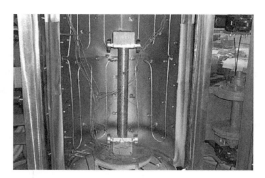

Figure 4.4 An X-band linac in a brazing furnace.

- The liquid brazing alloy does not penetrate into the cavity (a phenomenon known as *blushing*) and coat parts of the surface inside, which would change the resonant frequency of the cavity. Figure 4.5 shows an example of blushing.

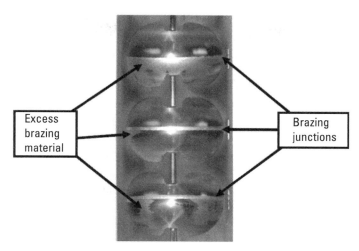

Figure 4.5 Excess braze material (blushing) on inner surfaces of cavities. (Courteously provided by Altair Technologies.)

It is to be noted that the cleaning processes mentioned above are critical in insuring that the joining surfaces are free of grease and oxide layers in order for the molten brazing filler to wet the joint during the brazing cycle. The brazing gap at the joint is typically 1 to 2 mils (a few hundredth of a millimeter). It is recommended that the brazed joint would be designed such that on cooling the joint is under some compression rather than in tension.

Different brazing alloys behave differently when becoming liquid and it requires experience and exact temperature control to attain the desired result. Some of the common brazing alloys used in Cu-Cu joints are listed in Table 4.2 [3]. The difference in brazing temperatures for different alloys allows the manufacturing engineer to design a process with multiple brazing cycles on the same component with successively reduced brazing temperatures. Actually, it is

Table 4.2
Common Brazing Alloys for Cu-Cu Braze and Their Typical Brazing Temperature

Brazing Alloy Composition	Typical Braze Temperature (°C)
35 Au/65 Cu	1,030
50 Au/50 Cu	990
20 Ag/60 Au/20 Cu	865
59 Ag/31 Cu/10 Pd	860
68 Ag/27 Cu/5 Pd	817
72 Au/28 Cu	790

From: [3].

possible, in some special cases, to do more than one braze at the same temperature using the same filler alloy. The reason we are able to do the successive braze with the same filler is that some fillers change their composition during brazing by alloying with the base material of the joining parts. Thus, the melting point of the new alloy moves to a value higher than the original melting temperature for the filler but is still lower than the melting point of the parts being joined. Therefore, the second braze would not melt the previous brazing.

The brazing is usually done in a dedicated furnace, which is either evacuated (vacuum of 10^{-6} torr or better) or filled with hydrogen. The vacuum braze helps in removing the hydrogen trapped in the linac parts and hence reduces the time needed to outgas the surfaces in the subsequent process of bake-out. In order to obtain uniform heating of large parts, the furnace must be equipped with a number of independent heating zones (typically molybdenum panels) whose temperatures are precisely controlled. The temperature control (by monitoring the readings from a sufficient number of thermocouples connected to the parts) must allow temperature homogeneity and precision within a few degrees. It is also possible to produce high-quality braze joints in a hydrogen atmosphere due to its reducing effect of the surface oxides. Hydrogen brazes help in getting more rapid and uniform heating of the components in the furnace. However, the hydrogen tends to remain in grain boundaries near the surface, which subsequently requires long bake-out to outgas this trapped hydrogen in order to attain good vacuum.

4.5.2 Diffusion Bonding

Brazing is used routinely in the assembly of linacs operating in the L (1–2 GHz), S (2–4 GHz), and C (4–8 GHz) frequency bands. For X-band (8–12 GHz) and higher frequency linacs, the ability to produce high quality brazed joints in a consistent way for small cavities has been a major concern. Excess flow of braze from the joints into the cavity leads to a local degradation of the surface finish and can produce an unpredictable change in cavity volume leading to frequency changes. These difficulties lead some accelerator research groups to develop the use of copper/copper diffusion bonding in the assembly of linacs [8, 9]. This process involves heating the mating surfaces in a furnace at about 1,000°C while the joining parts are kept under pressure. The contacting surfaces should be very clean and have superior flatness and surface finish. A good diffusion bond would be manifested by having the grains of one part grow into the other part across the bond joint at the joining surface, as shown in Figure 4.6. The research and development effort for about a decade, from the mid-1990s to the mid-2000s, helped in optimizing the conditions for diffusion bonding of linac copper cavities. In the early prototypes, the requirements on surface flatness and finish were very stringent, but after a decade of development these requirements

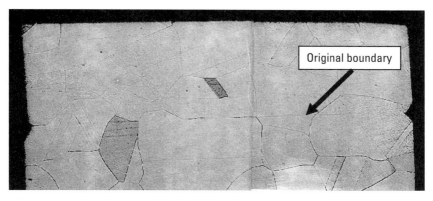

Figure 4.6 An example of a good bond between two OFE copper samples. (Personal communication, Chris Pearson, SLAC) [10].

were relaxed since it was found that we can increase the pressure exerted on the parts under the same heating conditions without the fear of introducing significant deformation to the linac cavities (resulting in detuning). In Table 4.3, diffusion bonding conditions at earlier linac prototypes are compared to more relaxed surface requirements after a decade of development.

In the diffusion bonding example shown in Figure 4.6, the OFE copper samples had a flatness of 200 μ" and surface finish of 8 μ", the pressure used was 20 psi and the temperature was 1,015°C (personal communication, Chris Pearson, SLAC).

4.6 Tuning of Linacs

During brazing, some of the molten brazing alloy can make its way inside the linac cavities, resulting in variation in the volume of these cavities, which in turn can change the resonant frequency characteristics of the linac. For this reason, it is a common practice to manually tune the individual cavities after the brazing step in order to bring the resonant frequencies of individual cavities

Table 4.3
Diffusion-Bonding Parameters

Surface Requirements and Diffusion Bonding Conditions	Mid 1990s	Mid 2000s
Surface Finish, Ra (μ")	0.4	8
Flatness (μ")	20	100
Temperature (°C)	1,015	1,015
Pressure (psi)	25	45

From personal communication, Chris Pearson, SLAC [10].

to their nominal design values. This is usually done by a skilled tuning technician who has to affix the linac on a fixture, perform a series of measurements [11], and modify the cavities as needed by deforming the physical structure of each cavity (dimpling it with a small hammer) until the desired frequency is achieved. The tuning can be done with the help of a computer program that uses explicit instructions and diagrams to lead the operator through a tuning sequence. The measurements are carried out using a Network Analyzer that is controlled by a computer using graphical programming environment such as LabVIEW©, a software by National Instruments [12]. Figure 4.7 shows, as an example, a computer screen for a computer-assisted tuning system for a side-coupled linac. In this example, the positions of measuring probes and auxiliary cavity inserts are displayed on the computer screen as a guide to the operator in tuning a specific cavity, the second cavity in this example. After the cavity is tuned, the program would display new positions corresponding to the next cavity to be tuned. The position indicators continue to move sequentially from one cavity to the next until the operator is done tuning all the linac's cavities. Such an automation approach of the low-power RF measurements for tuning proved to be effective in increasing the accuracy and consistency of the tuning process and at the same time reducing the tuning process cycle-time [11].

In order to be able to adjust the frequency of a linac cavity up and down, a small pin is sometimes brazed into the outer wall of the cavity. Then, a slide-hammer can be latched to the pin and used to "pull" or "push" the cavity wall, thus decreasing or increasing the cavity frequency, respectively. However, this

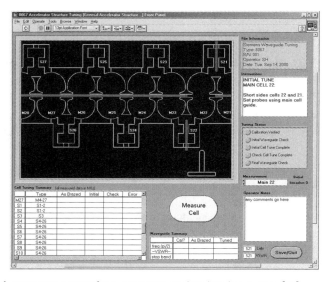

Figure 4.7 A computer screen for a computer-assisted tuning system [11].

additional provision of using a brazed tuning pin adds substantially to the cost and time of the linac production. Therefore, it is more common to machine the cavities to a slightly larger diameter and dimple the cavity wall to raise the frequency as needed.

As mentioned earlier in this chapter, copper work-hardens rapidly when distorted. For this reason, the tuning needs to be performed in as few dimples as possible and spread the dimpling around the surface of a cavity if possible. This is done because the work-hardening requires the use of greater and greater force on each successive dimpling at the same location.

It is to be pointed out that since the measurement of the cavity frequency is done in air under atmospheric pressure, the measured frequencies must be corrected for the dielectric constant of air and also for the ambient temperature, which may be different from the temperature at which the linac will be operated. For example, the dielectric constant of air at 20°C and 60% humidity is approximately 1.0006, which would lower the frequency by 0.3% from its vacuum value. The ambient temperature influences the dimensions of the linac, and hence, the frequency. As the temperature of the linac rises, the cavity volume is bigger and its frequency decreases. As a rule of thumb, the approximate change in frequency for S-band linacs (e.g., operating at 3 GHz) is 50 KHz/°C. For linacs operating in the X-band frequency range, such as those operating at 9 GHz, the decrease in linac's frequency would be 150 KHz/°C. The sensitivity of the linac's frequency to temperature varies depending on its design and the frequency band within which the linac operates.

4.7 Thermal Outgasing (Bake-Out)

The linac has to be sealed and kept under high vacuum for two reasons. First, the particles (electrons, protons, or ions) being accelerated should be able to travel free of any deflection caused by collision with gas atoms in the linac. Thus the interior space of a linac must have a low enough density of gas molecules to allow free passage of the accelerated particles. Second, the high electric field needed to accelerate the particle beam can ionize any remnant atoms or molecules floating inside the linac proper, resulting in arcing and breakdown. It is therefore desirable to maintain a vacuum level better than 10^{-6} torr. In fact, many linac manufacturers build accelerators with vacuum in the mid 10^{-8} torr scale. To achieve this level of vacuum in manufactured linacs, it is important to heat the linac inner surfaces while the linac is being evacuated. This process is called *thermal outgasing* or *bake-out*.

Normally, exposed copper surfaces would have a thin oxide layer. This porous layer can contain some gas species at concentrations more of than the Cu bulk. The rationale behind the bake-out process is based on providing thermal energy to gas molecules adsorbed on the surface or absorbed in the inner surface of the linac cavities to free these gases. In order to liberate these bound molecules, the thermal energy provided should be higher than the binding energy for each gas. This binding energy is exponentially dependant on temperature. Thus, the more we increase the copper temperature, the more outgasing we achieve. Typical bake-out temperatures for copper linacs would be in the range of 300°C to 500 °C. Typical gases that would be desorbed from the linac's inner surfaces are H_2, H_2O, CO, CO_2, and CH_4. It is also recommended that a device be attached to the system to analyze the type of gases that are being outgased from the inner surfaces of the linac. The device is called a residual gas analyzer (RGA), which contains a mass spectrometer for identifying different gas species. It is used to monitor the quality of the vacuum and easily detect minute traces of impurities in the low-pressure gas environment of the accelerator. It is a useful diagnostic device when the outgasing system does not achieve its nominal target vacuum level in the linac in the typical time duration for this process. The RGA can show specifically which gases are limiting the pump-down. This enables the accelerator manufacturing engineer to trace the source of the excess gas or contaminants present and take the appropriate corrective action in order to return the linac manufacturing processes to normal operation.

Bake-out is a slow process that has a relatively longer cycle-time in the linac manufacturing flow compared to other processes. It may take 20 to 30 hours to attain vacuum levels of 10^{-8} to 10^{-9} torr. One way to optimize the bake-out cycle-time is to end the process based on attaining a certain vacuum level that is sustained for a specific period of time. This is a better approach than just assigning a certain fixed duration for the outgasing process. Also, some outgasing furnaces can be built to accommodate two or three linacs to be processed at the same time, increasing the throughput of the process.

After the linac has been evacuated to the desired vacuum level, the linac is disconnected from the pumping system of the bake-out station. The connecting vacuum port is flanged, sealed, or pinched off. Now, the vacuum level of the linac is sustained with a small vacuum pump mounted on the linac. Typically this is an ion pump, since these pumps are the appropriate type of pumps for low-pressure systems such as the linacs vacuum systems.

4.8 Electron Gun Activation

The next step in the linac manufacturing progression is to prepare the electron gun for emitting a steady flow of electrons. Usually, this process is called *gun activation* or *gun conditioning*. As mentioned in Chapter 2, the electron gun is the part of the linac that supplies the stream of electrons, usually as a well-defined beam. It consists of a cathode, a heating filament, and a grid assembly.

In many cases, the anode for the gun is the linac copper body. Specifically, this step involves activating the cathode of the electron gun by heating it by means of applying a voltage across the heating filament. The heating is done in gradual steps at the beginning of the process to allow the linac's ion pump to pump away gases breaking away from the cathode and the internal surfaces of the electron gun. The goal of this cathode activation is to promote the formation of a layer on the cathode surface that emits electrons easily and steadily during the linac's normal operation. By applying heat during gun activation, the chemicals impregnated in the cathode tungsten body diffuse to the emission surface to form a uniform layer that has a low work function, making it a good electron emitter. The end of the gun activation process is usually verified by plotting the gun emission curves. These curves are plots of the measured current emitted versus filament voltage (which is proportional to the cathode temperature), as simulated in Figure 4.8. From these plots we can differentiate between a fully activated gun (Plot 1) and a partially activated gun (Plot 2). A fully activated gun would exhibit a flat region with constant emission, called the space-charge region. Operating in this region ensures the stability of the current emitted against variations in the cathode's temperature.

Some of the pitfalls inherent in gun activation when done improperly are as follows:

- *Incomplete gun activation.* This results in an uneven emission from different regions on the cathode surface. This would be manifested in the gun emission plots as the Plot 2 in Figure 4.8. Here we notice that we do not get the nice flat region, and hence we lose the stable operation of the gun.

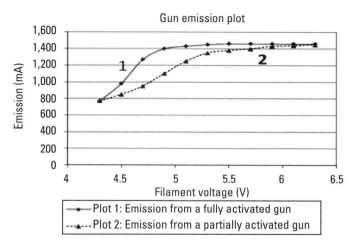

Figure 4.8 Simulated emission plots for an electron gun.

- *Overheating the cathode.* This would cause loss of the impregnation emission material prematurely through evaporation. This can also result in the deposition of some of this emission material on linac surfaces, with the accompanying lowering of the work function of the linac copper surfaces coated with this material. These linac surfaces would now emit electrons under the forces of the high RF fields during normal operation. The adverse effect of that is manifested as a current emitted in the linac even when the electron gun is off. We call this undesirable current *dark current*.

4.9 High-Power RF Conditioning

Before we can begin supplying the full RF power into our linac we need to condition the inner surfaces of the accelerator so that it would be able to stand high fields without arcing or breaking down. This conditioning step is done by increasing the RF power input to the linac slowly and gradually. This process is usually known as RF conditioning or RF processing.

A newly fabricated linac will break down (i.e., experience arcing) frequently at high accelerating fields. The presence of high electric fields near the copper surface can give sufficient energy to electrons in the copper that are near its surface to escape from that surface. This phenomenon is known as high-field emission. The freed electrons then get accelerated by the RF fields and can collide and ionize atoms or molecules floating inside the linac, resulting in arcing. As the linac is operated, the arcing rate at a given field strength decreases gradually. When we increase the field strength or the RF pulse width to new higher values, the arcing rate increases again. This cycle of RF processing is repeated until at some point no further significant progress (reduction in the arcing rate) can be made in a reasonable time of processing. The simple explanation of the mechanism of conditioning is that RF processing vaporizes, or "polishes away," small surface features or whiskers on the inner surfaces of the linac. Actually, in the process, some molten metal splashes from the vaporization spot to nearby ones, forming new features. The size of the new feature depends on the strength of the RF field in the linac during processing. For this reason, it is important that input RF power would be increased *gradually and in small increments* to avoid the creation of new features that would cause further arcing. Increasing the pulse width should also be done very carefully, since the pulse width determines the amount of RF energy available to sustain an arc once it is initiated.

Generally speaking, the above procedure is a slow and lengthy process. For this reason, it is recommended to automate the process whenever feasible. An example for the conceptual algorithm behind a linac automated conditioning system is described in the publication [13] and depicted in Figure 4.9. In this technique, three parameters are controlled, namely:

- Peak RF power;
- Pulse repetition frequency (PRF);
- Pulse width.

These parameters are stepped up automatically in increments until reaching the maximum power rating for the specific linac being conditioned. This is done with the help a feedback system that continually monitors two feedback parameters: arc counts and ion pump current (indicating the vacuum level).

It is to be noted that some of the linac manufacturing processes preceding the conditioning step impact the length of time needed to fully RF condition a linac. Two of these factors are:

- Scratches, machining marks, or sharp features on the inner surface of the linac. These act as points of field enhancement, and hence they become more susceptible to high-field electron emission.
- Impurities and contaminants on the inner surface of the linac. These can emit electrons under the effect of high field easier than the copper (i.e., they may have a lower work function compared with pure copper).

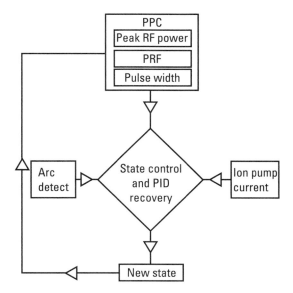

Figure 4.9 Conceptual algorithm for a linac automated conditioning system [12].

From the above, it is clear that a linac process engineer needs to pay attention to the machining process of the copper to ensure that the inner surfaces are free from scratches, visible machining marks, or sharp features. Also he or she should carefully guard the cleanliness of the linac components through all steps of the linac manufacturing.

4.10   Linac's Beam Tests and Test Bunkers

After the linac has been conditioned to operate under high RF power, it goes into a set of beam tests to ensure that it meets its operational specifications. The linac is tested under conditions simulating its normal operation in the field (whether in a cancer treatment unit or in some industrial application). The output electron (or X-ray) beam is tested for its energy, dose rate, and stability. The input RF power needed to achieve the specified output is also measured. In certain linac applications it is important to identify and record the beam's transverse profile. The results of the beam tests should be documented and sent to the user of the linac with other documentations accompanying the shipped linac.

The linac conditioning and beam tests are carried out in a specially shielded room. The goal is to limit radiation exposures to acceptable levels for the test operator and the supporting technicians by running the linac tests from a control console outside this *test cell* or *bunker*. The shielding material absorbs the radiation according to the density of material. Concrete is usually the material of choice, providing a good compromise between cost and effective radiation absorption. However, if space is at a premium, then special high-density concrete or high-density metals, such as steel or lead, can be used. The thicknesses of the concrete should be commensurate with the maximum energy of the beam used during testing. To guard against air pockets, it is customary to vibrate the concrete mix as it is poured. Ideally, the concrete should be formed in one pour to avoid seams between different layers. The test bunkers are usually located on the periphery of the linac manufacturing complex to avoid radiation exposure to other workers located in high-occupancy areas. Actually, in some facilities, the linac testing is done in rooms below ground level to save on shielding costs and reduce radiation leakage to the surrounding working areas.

In order to reduce the radiation dose near the entrance, a restricted access passageway leading to the test bunker may be incorporated in its layout. This passageway is termed the maze (see Figure 4. 10). Another advantage of a maze is a route for ventilation ducts and electrical conduits without compromising the shielding walls.

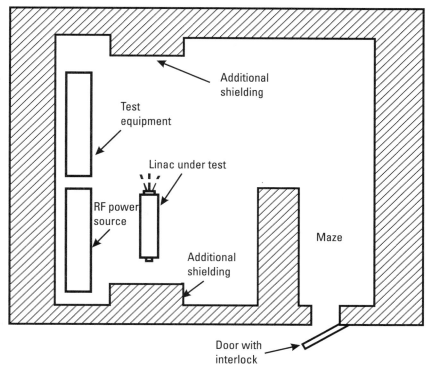

Figure 4.10 A typical layout for a linac test bunker.

The door to the test bunker must be interlocked to the accelerator test setup, such that the irradiation will be terminated to prevent an exposure if a door is inadvertently opened. The interlock should be fail-safe so that safety is not jeopardized in the event of failure of any one component of the system. Also, it is recommended to have an emergency beam turn-off switch installed in appropriate location in the accelerator test room.

The test control area, where the operators control the linac tests, should be close to the entrance to the test bunker so that the operators can view the entrance area. It is also recommended that an illuminated warning sign be displayed at the entrance to the test bunker, which is illuminated when the beam is turned on and delivering radiation.

It should be noticed that ducts and conduits between the test cell and the outside control console must be adequately shielded to prevent any radiation leakage through them. This includes ducts for cables necessary to control the test unit and physics equipment, as well as heating and ventilation ducts.

4.11 Common Manufacturing Issues and Imperfections

We are now essentially done with our guided tour of the linac manufacturing facility. Nevertheless, as many process engineers know, it is expected and accepted that the above linac manufacturing processes inherently and inevitably encompass some degree of deviation from the nominal process specifications. For those working in this industry, I cannot overemphasize the importance of cleanness of the parts and the manufacturing environment. Also important is having a quality system in place to avoid the common linac manufacturing issues and potential defects. Therefore, it is the role of the process engineer to:

- Understand the predictable manufacturing imperfections and systematic variations in each process and the impact on subsequent processes and on the performance of the final product, the linac.
- Implement various quality system (QS) provisions such as quality control (QC), quality assurance (QA), and statistical process control (SPC), and integrate them into different linac manufacturing processes.
- Continuously implement different process improvements. The process engineer can always benefit from measures and approaches derived from the Six Sigma methodology [13].

In this section, I list some of the linac manufacturing processes' imperfections and deficiencies and their typical impact on the quality of the linac. These are tabulated in Table 4.4. Quality measures and some examples of QA, QC, and SPC relevant to linac manufacturing are discussed in the next section.

4.12 Quality Systems in Linac Manufacturing

Many of the commercial linacs are used as part of radiation therapy machines used in treating cancer. Consistent high quality and defect-free products are clear requirements in the field of medical devices. Manufacturing reliable accelerators and, hence, reliable radiation therapy machines requires accurate, efficient, and well-controlled processes throughout the steps of the accelerator's manufacturing. Thus, the linac manufacturing processes normally encompass built-in QSs that are applicable to the general medical device industry, a heavily regulated industry. In the United States, these processes benefit from the guidelines offered by the U.S. Food and Drug Administration (FDA). In this section, I highlight briefly some of the built-in QC and QA measures as well as SPCs that can be implemented in the manufacturing of medical and industrial linacs. These measures should result in processes that are stable, predictable,

Table 4.4
Potential Linac Manufacturing Imperfections

Manufacturing Imperfection	Impact on Linac Operation or Quality of Manufacturing
Visible machining marks or scratches on the linac inner surface	Increase in arcing → Less efficient linac
Contaminated machining coolant or lubricant	Impurities embedded inside the copper surface → Increase in outgasing time Increase in RF conditioning time Increase in arcing
Incorrect cavity dimensions	Lower or higher frequency → Less efficient linac
Incomplete chemical cleaning	Contaminated copper surface → Increase in outgasing time Increase in RF conditioning time Increase in arcing
Imperfect alignment of cavities	Less efficient linac Lower output dose-rate
Excess brazing material inside the linac	Possible effects: Increase in outgasing time Increase in RF conditioning time Increase in arcing Difficulty in tuning and possible shift in the linac's operating frequency Increase in linac's wall loss → Less efficient linac
Having pockets of trapped gas (virtual leak)	Increase in outgasing time
Error in tuning a cavity or more	Shift in the linac operating frequency Less efficient linac
Misaligned electron gun	Less efficient linac Lower output dose-rate
Excessive heat in welding	Change in frequency of adjacent cavities
Incomplete thermal outgasing	Increase in RF conditioning time Increase in arcing
Incomplete gun activation	Unstable beam output
Incomplete RF conditioning	Increase in arcing Missed pulses in beam output

and guarantee the production of linac-based machines that are of high quality, cost-effective, and in compliance with regulatory requirements.

4.12.1 Examples of QC and QA Measures

4.12.1.1 Cell Machining QC

Because the linac cavities host very high fields during the linac operation, and to avoid arcing and breakdown, the cavity's inner surfaces should be free of scratches. For this reason, it is not recommended to verify the dimensions of the machined copper cells using a stylus or a tool that would touch the inner surface of the cells. An alternative technique to confirm that the cell is machined to specifications is to verify the frequency of the machined cells [15, 16]. A typical set-up would be based on a computer-controlled network analyzer to measure the resonant frequencies of individual cells as they come out of the lathe. In this setup, the cell is placed on a clean metallic plate that acts as a conducting wall to close up the cell to form a resonant cavity (Figure 4.11). The plate has holes where RF probes are protruding to excite and pick up the signal inside the cavity. The feedback from this technique is used to "tune" the lathe for each batch and also periodically during the day to compensate for tool wear or temperature variations. The frequency of each cell and its serial number are recorded.

4.12.1.2 Bead-Pull Technique as a QA Measure

Tuning a linac would confirm that the frequencies of the linac cavities are at the correct value. However, it is also important to make sure that the electric field distribution inside the linac is as specified by the linac design. This is done using a perturbation technique known as bead-pull [17]. This measurement technique is based on the fact that a small object (a bead), when placed inside any of the linac cavities, will change the energy stored in the electric field in this cavity. This is because the bead has dielectric properties different from those of air and thus perturbs the electric field at its position. This results in a change in the resonance of this cavity. A simplified, and actually an approximate, view of the effect of the bead is that its presence changes in the capacitance of the cavity under the assumption that the fields outside the bead are hardly altered. The change in cavity capacitance means a change in the electrical energy stored in the cavity and at the same time a change in the cavity frequency. Thus, the relative change in frequency is proportional to the relative change in the electric

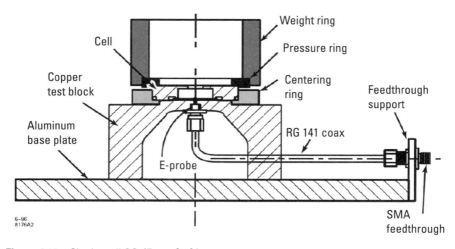

Figure 4.11 Single-cell QC. (From: [15].)

energy stored in a cavity. Since the electric energy is proportional to the square of the electric field, we can construe that the strength of the electric field can be plotted as a function of position along the linac by detecting the change in frequency (Δf) of the cavities as the bead moves along the linac. If we call the unperturbed energy stored in the cavity U and the unperturbed electric field E, then we can describe the concept behind the bead-pull mathematically as:

Relative change in frequency $\xrightarrow{\text{is proportional to}}$ relative change in energy stored. Stored Energy $\xrightarrow{\text{is proportional to}}$ the square of the electric field.
$$\Delta f / f_0 \propto \Delta U / U$$
$$E \propto \sqrt{U}$$

Thus, $E \propto \sqrt{\Delta f}$

Or, $E(\text{position}) = \text{Constant} \times \sqrt{\Delta f}(\text{position})$.

The electric field amplitude distribution for a standing-wave accelerator resulting from a bead-pull measurement is portrayed in Figure 4.12.

A typical bead-pull setup uses a network analyzer to measure the change in frequency as a function of the position of the bead in the linac. A small bead is normally affixed to a thread that moves inside the linac with the help of a system of pulleys and a step motor. Figure 4.13 shows a simplified schematic for a bead-pull system [18]. Note that the linac shown in this setup is of the traveling wave type and the reader may notice that this linac, as with all TW linacs, has two RF ports, an input port and an output port.

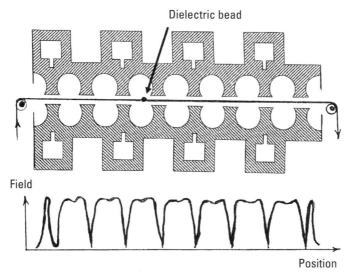

Figure 4.12 A simplified representation of an electric field distribution from a bead-pull scan of a standing-wave linac.

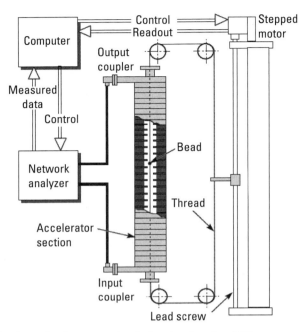

Figure 4.13 A schematic of an example of a bead-pull system [17].

4.12.2 Examples of Statistical Process Control

One of the techniques that can be used to monitor different accelerator manufacturing processes and provide useful feedback to the process engineer is SPC [19]. Analysis of the data resulting from the SPC is used to initiate actions to maintain and improve the capability of the process. SPC is used to control the process by signaling when adjustments may be necessary, thus preserving consistent and high-quality process performance. SPC uses control charts as a graphical means for monitoring critical process parameters. A control chart allows the operator to monitor trends occurring in a process by plotting the values of a chosen process parameter as a function of time or sample number. A process is in "a state of control" if this plotted parameter lies between the upper and lower control limits of the process. The upper control limit (UCL) and the lower control limit (LCL) are determined by evaluating the dispersion (variability) in process (see Figure 4.14). In a well-controlled process, these limits can be chosen to be equal to $\mu \pm 3\sigma$ respectively, where σ (sigma) is the process standard deviation and μ is the process mean. These statistical limits are normally called the 3-sigma control limits. In a normal (Gaussian) distribution, 99.73% of the parameter-tracked values lie in an interval of width 6-sigma.

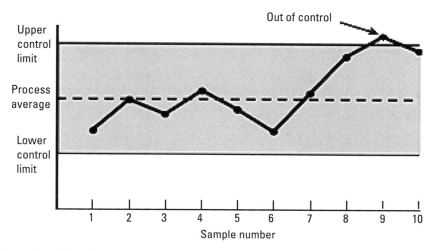

Figure 4.14 General form of a control chart for SPC.

4.12.2.1 Possible SPC Parameters in the Manufacturing Processes of a Linac

As a guide to the linac process engineer, I list below some of the linac manufacturing parameters that can potentially be good candidates for SPC tracking:

> *Receiving inspection parameters.* Tracking some of critical parameters of the received components and subsystems is important for evaluating and ensuring the supplier's quality.
>
> *Cavity frequency "as-machined."* This is a measure of the consistency of machining and also helps detect any adverse trends (such as tool wear or lathe misalignments).
>
> *Cavity frequency "as-brazed."* This parameter reflects the consistency of the brazing process. For example if there is internal blushing (excess braze material), it would be manifested as an increase in the cavity frequency.
>
> *Tuning parameters.* These parameters indicate the overall quality of tuning the accelerator.
>
> *Bake-out time.* The bake-out process time needed to achieve the specified vacuum level, as indication of completion of the process, is monitored and recorded as an SPC parameter. This parameter can be used as an indication of the level of cleanliness of the preceding processes. If an RGA is attached to the bake-out station, one could also track the level of a particular gas at a particular point in time, say at the beginning or end of bake-out.

E-gun activation time. This would allow the process engineer to track the consistency of the performance of the electron guns and, hence, the consistency of the manufacturing processes at the E-gun supplier.

RF conditioning time. This parameter can be used as an indication of the level of cleanliness of the preceding processes as well as internal surface finish.

In Figure 4.15, I show a flow chart for linac manufacturing and mark possible processes for which tracking their characteristic parameters by SPC can be implemented.

4.13 Guidelines for Linac Buyers and Users

In addition to reviewing linac's manufacturing processes, we also discussed in this chapter the linac manufacturing processes' imperfections and deficiencies and their typical impact on the quality of the linac. It is hoped that this would help buyers of linacs for medical or industrial use understand explanations given by linac manufacturers about why a particular linac may not be performing up to its full specifications.

Another goal of this chapter is to help potential linac buyers and users in assessing the merits of alternative linac suppliers. In choosing a linac supplier, one can consider the following list of guiding remarks and points:

1. Records for QC and QA data provided by the linac supplier.
2. The extent and consistency of tracking quality parameters of the manufacturing processes. A good linac manufacturer would implement a quality-tracking system, such as SPC tracking.
3. Examination of the material specifications and certifications indicating the type of copper used and its purity and chemical contents.
4. Cleanliness practices in handling and transporting linac parts prior to the linac final assembly.
5. Indications of excessive tuning, such as a disproportionate number of tuning dents and markings.
6. Excessive overflow of brazing material on the external surfaces of the linac (blushing). As mentioned in Section 4.6 of this chapter, this would indicate the existence of internal blushing, which results in changes in the resonant frequencies of the linac cavities.
7. Indications of repeated or excessive welding. This can result in changes in frequency of cavities bordering the weld region. Normally, it is dif-

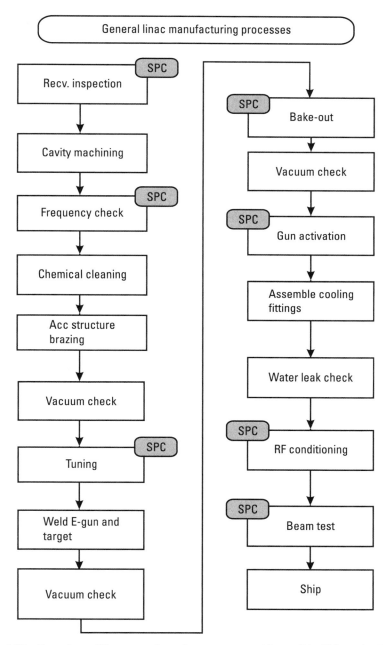

Figure 4.15 Flow chart of linac manufacturing processes with possible SPC monitoring.

ficult to measure and correct the frequency variation in an assembled linac subsequent to the tuning process and sealing the linac.

8. Checking the vacuum integrity of the linac. Poor vacuum in the linac can be indicated by higher than normal current reading on the ion-pump power supply.
9. Excessive arcing as indication of less-than-perfect cleaning process, incomplete thermal out-gazing, or incomplete RF conditioning.
10. Some linac manufacturers prefer to use bolted flanges to mount the electron gun. Others avoid flanges and use welding to mount the electron gun. The former approach has the advantage of giving the manufacturer's technicians the ability to change the electron gun in the field. The later approach avoids any risk of compromising the vacuum integrity by breaking vacuum in the field. However, it requires sending the linac back to the manufacturer for electron gun replacement.

References

[1] Pekeler, M., et al., "Fabrication of Superconducting Cavities for SNS," *Proceedings of LINAC 2004*, Lübeck, Germany, 2004, pp.602–604.

[2] Padamsee, H. et al., *RF Superconductivity for Accelerators*, John Wiley and Sons Inc., 1998, pp. 105–127.

[3] Wilson, I., "Cavity Construction Techniques," *CAS CERN*, Vol. 92-03, 1992, pp. 375–349.

[4] Wilson, I., et al., "The Fabrication of a Prototype 30 GHz Accelerating Section for CERN Linear Collider Studies," *Proceedings of the 1990 European Particle Accelerator Conference (EPAC)*, 1990, pp. 943–945.

[5] Mathewson, A. G., "Making it Work," *CAS CERN*, Vol. 92-03, 1992, p. 421.

[6] Arkan, T., et al., "Fabrication of X-band Accelerating Structures at Fermi Lab," *Proceedings of the 9th European Particle Accelerator Conference*, Lucerne, Switzerland, July 2004, pp. 815–817.

[7] Lafferty, J. M. (ed.), *Foundations of Vacuum Science and Technology*, John Wiley and Sons Inc. 1998, pp. 487–493.

[8] Elmer, J. W., et al., "Diffusion Bonding and Brazing of High Purity Copper for Linear Collider Accelerator Structures," *Phys. Rev. ST Accel. Beams*, Vol. 4, 053502, 2001, pp. 1–16.

[9] Bagnato, O. R., et al., "Development of Diffusion Bonding Joints Between Oxygen Free Copper and AISI 316L Stainless steel for Accelerator Components," *Proceedings of IPAC'10*, Kyoto, Japan, May 2010, pp. 3975–3977.

[10] Pearson, C., private communication, SLAC National Accelerator Laboratory, Menlo Park, CA.

[11] Hanna, S. M., "Characterization Techniques for X-Band Medical Accelerator Structures," *Proc. of EPAC 2000*, Vienna, Austria, 2000, pp. 2521–2523.

[12] National Instruments company Web site, "LabVIEW System Design Software," www.ni.com/labview/.

[13] Hanna, S. M., and S. Storm, "Automated High Power Conditioning of Medical Accelerators," *Proceeding of the 9th European Particle Accelerator Conference,* Lucerne, Switzerland, July 2004, pp. 2795–2797.

[14] Bicheno, J., and P. Catherwood, "Six Sigma and the Quality Toolbox: For Service and Manufacturing," *Buckinghamshire,* UK: Picsie Books, 2005.

[15] Hanna, S. M., et al., "Microwave Cold-Testing for the NLC," *Proc. of the Fifth European Particle Accelerator Conf.,* 1996, pp. 2056–2058.

[16] Hanna, S. M., et al., "Development of Characterization Techniques for X-Band Accelerator Structures," *Proc. of the 1997 IEEE Particle Accelerator Conference,* 1997, pp. 539–541.

[17] Wangler, T. P., *Principles of RF Linear Accelerators,* New York: Wiley, 1998, pp. 163–164.

[18] Hanna, S. M., et al., "Semi-Automated System for the Characterization of NLC Accelerating Structures," *Proc. of 16th IEEE Particle Accelerator Conference (PAC 95),* 1995, pp. 1108–1111.

[19] Levinson, W. A., and F. Tumbelty, *SPC Essentials and Productivity Improvement: A Manufacturing Approach,* Milwaukee, WI: ASQC Quality Press, 1997.

5

Role of Linear Accelerators in Cancer Radiation Therapy

5.1 Basic Radiation Therapy Concepts and Definitions

Radiation therapy or *radiotherapy* (RT) describes the clinical process that uses ionizing radiation to kill cancerous cells. It is one of the major medical modalities used in treating cancer. Other widely used cancer treating modalities are surgery and chemotherapy. Actually, more than half of all cancer patients receive radiotherapy, either alone or in combination with surgery or chemotherapy. The power of the radiation therapy is its ability to ionize atoms and molecules inside the nuclei of the biological cells of the tissue to which the radiation is applied, thus killing the cancerous cells by damaging their DNA. The radiation can either damage the DNA directly or create charged particles (free radicals) within the cells that can in turn damage the DNA.

A patient may receive radiation therapy before, during, or after surgery, depending on the type of cancer being treated. Also, some patients may receive radiation therapy in combination with chemotherapy. In certain cases both radiation therapy and chemotherapy are used after surgery. In a few number of cases, other cancer treatments are used, such as those employing nonionizing radiation. Examples are the photodynamic (light) therapy or hyperthermia (heat) treatment.

Depending on the stage of cancer, RT is used either as a curative treatment or as a palliative treatment, relieving some of the symptoms of the disease and reducing the suffering caused by it, thus improving the quality of life of cancer patients. The ultimate goal of radiation therapy is to deliver an accurate dose of ionizing radiation to a well-defined targeted tumor volume and, at the same time, satisfy the challenge of minimizing the dose of radiation absorbed by the surrounding healthy tissue.

In general, radiation is classified into two main categories, nonionizing and ionizing. The distinction is based on the ability of the radiation to ionize the absorbing material (see Figure 5.1). Ionization is the ejection of an orbital electron from the absorber atom. Medical (and industrial) applications of linacs utilize the ionizing radiation. Ionizing radiation can ionize matter either directly or indirectly:

- Directly ionizing radiation are charged particles: electrons, protons, and ions. They deposit energy in the medium through direct Coulomb interactions between the charged particles and orbital electrons of atoms in the medium.
- Indirectly ionizing radiation are photons (X-rays and gamma rays) and neutrons. The effect of this type of radiation is done in two steps. First, the incident radiation creates charged particles: photons release electrons or positrons, and neutrons release protons or ions. Second, the released

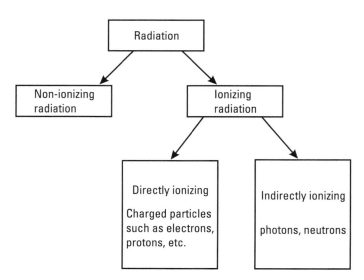

Figure 5.1 Categorization of types of radiations.

charged particles deposit energy to the medium through direct Coulomb interactions with orbital electrons of the atoms in the medium.

Most radiotherapy treatments are carried out with photon beams; however, some are carried out with electron beams. Electron beams are especially used for superficial tumors (less than 5 cm deep). For this reason, many of the linac-based RT machines provide the option of using photon beams as well as electron beams for cancer treatments.

Ionizing radiations used in conventional linac-based RT units are high-energy electron beams, and X-rays, and are the subject of this chapter. Radiation therapy techniques using protons and carbon ions have some unique characteristics but require higher energy accelerators. They are based on circular accelerators, such as cyclotrons or synchrotrons, and hence this type of RT is offered in larger facilities, some of which are discussed in Chapter 7.

X-rays and gamma rays are electromagnetic waves. The discrete quantity of electromagnetic radiant energy is the *photon*. When a photon collides with an atom's orbital electron, it knocks this electron out of position, ionizing the atom. These newly freed electrons can ionize other atoms and molecules in the irradiated tissue. Although one cannot distinguish between X-rays and gamma rays if they both have the same energy, it is customary to apply the term *X-ray to* ionizing electromagnetic radiation produced by machines. The term *gamma rays* is applied to the similar radiation produced by radioactive isotopes (*radionuclides*).

The parameter that quantifies the biological impact of ionizing radiation on irradiated tissues is the *absorbed dose* or simply *dose*. The unit of the dose is the *rad* (which is derived from *rad*iation *a*bsorbed *d*ose). It is equivalent to the deposition of 100 ergs of energy per a gram of irradiated material. The SI unit of dose is the *gray* (Gy), which is the absorption of one joule (J) of energy per a kilogram of absorbing material. The *dose rate* is specified in rads per minute or centigray per minute.

Recalling that, 1 joule = 10^4 ergs, then we get:

1 rad = 100 ergs/g = 10^{-2} J/kg
1 Gy = 1 J/kg
1 rad = 10^{-2} Gy = 1 cGy

Thus, one rad is equal to one *centigray* (cGy)

Note that because the biological effects of radiation depend not only on dose but also on the type of radiation, the parameter used in dosimetry relevant to radiation protection is the *dose equivalent*. The unit for the dose equivalent is the sievert (Sv). To take into consideration the type of radiation, the dose is multiplied by a dimensionless weighing factor W to get the dose equivalent. One sievert is equal to one gray times the factor W.

$$1 \text{ Sv} = 1 \text{ (Gy)} * W$$

For X-rays, gamma rays, and electrons this weighing factor W is one. Thus for X-rays, gamma rays, and electrons, we can treat the Sv as one Gy, or 100 rads. The factor W takes larger values for radiation of heavier particles, such as protons and carbon ions.

The dose equivalent is commonly expressed in units smaller than the Sv. In the United States, the units *rem* and *mrem* (millirem) are used frequently.

1 rem = 0.01 Sv = 10 mSv

1 mrem = 0.001 rem = 0.01 mSv = 10 μSv

When a beam of electrons passes through a medium, the electrons suffer multiple *collisions* with orbital electrons as well as change in their direction of travel (i.e., *scattering*) due to Coulomb force interactions between the incident electrons and the nuclei of the medium. These two effects cause the electron beam to lose some of its kinetic energy as it travels through the medium. Thus, for an incident electron beam of electrons, the dose is reduced with depth. Similarly, a photon beam propagating through a medium is affected by attenuation and scattering of the photon beam inside the medium. The variation of dose with depth is normally described by the beam percentage depth dose (PDD). It is the ratio of the dose at a given point on the central axis of a beam (electron

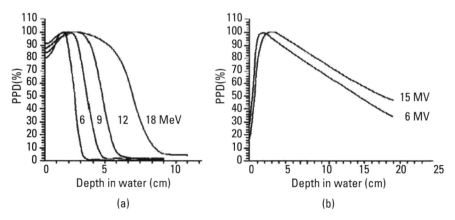

Figure 5.2 Typical PDD curves in water for (a) electron beams with energies 6, 9, 12, and 18 MeV and (b) photon beams with 6 and 15 MV [1].

or X-ray) to the maximum dose on the central axis, multiplied by 100. Typical PDD curves in water for both electrons and X-rays are shown in Figure 5.2.

5.2 Radionuclides-Based Radiation Therapy

There are two principal approaches for delivering ionizing radiation to the tumor. In the *internal radiotherapy*, a radiation source is placed inside or in a close proximity to the area of the targeted tumor. This approach is known as *brachytherapy*. This is in contrast to *external beam radiotherapy*, also known as *teletherapy*, in which the ionizing radiation originates from a machine outside the body and is directed towards the tumor.

5.2.1 Brachytherapy

Brachytherapy is based on placing a radioactive source in close proximity to the tumor. In this mode of therapy, a high dose of radiation can be delivered locally to the tumor with rapid dose fall-off in the surrounding healthy tissue. Most of the sources used in brachytherapy are artificially produced radioactive isotopes such as cesium 137 (^{137}Cs), iridium 192 (^{92}Ir), gold 198 (^{198}Au), iodine 125 (^{125}I), and palladium 103 (^{103}Pd). The implanted radionuclides can be removed after a period of time (*temporary implant*) or remain forever (*permanent implant*). Brachytherapy is used alone or combined with external beam radiotherapy. About 5% to 15% of patients receiving radiation may be candidates for brachytherapy. Further coverage of brachytherapy is beyond the scope of this book. Readers interested in more information will find numerous books on the subject [2–4].

5.2.2 Cobalt Teletherapy

In this category of external beam radiotherapy, gamma rays are emitted from radionuclides as they undergo radioactivity disintegration. The isotope cobalt 60 (^{60}Co) proved to be the most suitable radionuclide to be employed in this type of RT machines. The ^{60}Co emits gamma rays, which constitute the useful treatment beam.

The isotope ^{60}Co is produced when cobalt 59 (^{59}Co) is placed in a nuclear reactor in which the nucleus captures a neutron to become unstable and hence turn into a radioactive isotope. The isotope ^{60}Co has a half-life of 5.27 years and emits two gamma rays with energies of 1.17 and 1.33 MeV, corresponding to an average energy of 1.25 MeV.

The ^{60}Co source, which is usually in a form of a solid cylinder, discs, or pellets, is typically sealed inside stainless-steel capsules housed inside a steel shell with lead for shielding purposes. The housing of the radioactive source is a part of a gantry to facilitate the rotation of the source about a horizontal axis, referred to as the machine *isocenter axis*. In addition to the source housing, the gantry comprises a beam collimator and source movement mechanism for bringing the source in front of the collimator opening to produce the clinical gamma-ray beam. A cobalt teletherapy unit is shown in Figure 5.3 [5, 6].

The cobalt unit has the advantage of reliability because of the relatively straightforward configuration. However, the margins of the gamma-ray beam have a less well-defined border than the X-ray beams from linacs. The lack of definition of the beam margin, known as the *penumbra*, is the result of the relatively large physical size of the cobalt source, which is of the order of 2 cm in diameter, compared with RT units using linacs that have a focal spot size of about 1–2 mm. The closer the radiation source's size to a point source, the smaller the penumbra. The larger penumbra results in fuzziness at the edges of the radiation field and, hence, the possibility of a higher level of irradiation of tissues adjacent to the treated volume. This can be an important factor in limiting the discrimination between the targeted tumor and the surrounding healthy tissue in delivering the radiation using the cobalt units. Another problem with

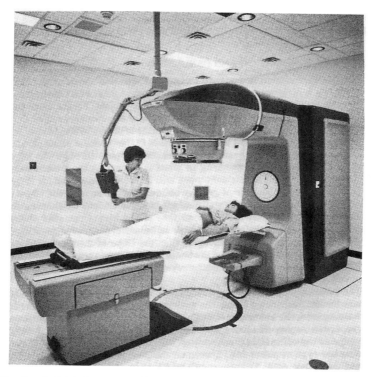

Figure 5.3 A cobalt teletherapy unit (Theratron 780). (Courtesy of Atomic Energy of Canada, Ltd. now MDS Nordion.)

cobalt units is their decaying source, which results in gradual reduction in output. This in turn results in progressively increasing treatment times with the associated continued reduction in patient throughput. Generally, the cobalt source needs to be replaced every 5 to 7 years, with the accompanying cost and additional provisions needed for the disposal of the spent source.

Today, the overall use of such radionuclide-based treatment units has decreased in many countries around the world and many have been replaced with linac-based radiation therapy units. Some of the reasons for this decline are mentioned above, but perhaps the most important reason is the risk associated with the storage of radioactive sources in a hospital and the disposal of the spent cobalt. In contrast, medical linacs contain no inherently radioactive materials.

5.3 Accelerator-Based Radiation Therapy

The development of both the cobalt-based teletherapy units and the linac-based radiation therapy machines started roughly at the same time in the midfifties of the last century. However, the linac-based RT machines eventually eclipsed the cobalt units and became the most widely used radiation source in the present day. Linear accelerators can produce higher energy beams than cobalt units and operate at higher dose rates. With their compact and efficient designs, linac machines offer excellent versatility with a wide range of energies. They can provide either electron or X-ray therapies with megavoltage beam energies.

In addition to the conventional radiation of electrons and X-rays, accelerators can be used to accelerate heavier particles, such as protons and ions of certain elements, to treat cancer. This recent effort to develop *particle therapy* has produced several treatment centers around the world. Particle therapy continues to benefit from a rigorous effort of research and development to make it a viable and an affordable cancer therapy option. We will cover particle therapy in Chapter 7.

In the conventional therapy, the majority of treatments use X-rays and a smaller number uses the electron beam or a combination of both therapies. In the *X-ray therapy*, they can produce photon beams in the range of: 4 to 25 MV, and in the *electron therapy* the treatment electron beam can cover the range of: 6 to 25 MeV. The length of the linac depends on the final electron kinetic energy, and ranges from ~30 cm for 4 to 6 MeV linacs to ~150 cm for the 25-MeV linacs. The linac is mounted on a gantry. As the gantry is rotated, so is the accelerator and the resulting treatment beam, so that the radiation can be delivered to the tumor from multiple directions. In Figure 5.4 we show one of Siemens' radiation therapy machines, the Primus™. This radiation oncology machine offers both X-ray and electron beams and at multiple energies. It uses the longer type linac that is located in the gantry parallel to the gantry's axis of

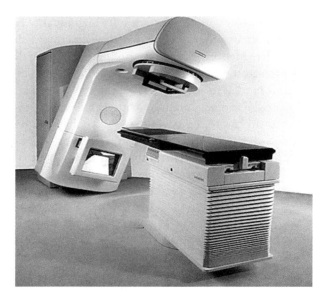

Figure 5.4 Example of an accelerator-based radiation therapy machine, the Primus ™, by Siemens. (*From:* [7].)

rotation. A beam transport system is then used to transport the electron beam from the accelerator to the treatment head including a 270° bending magnet.

5.4 The Medical Linac Requirements

The electron linear accelerator is the source of the ionizing radiation in the prevailing majority of radiation therapy units. It can be a standing-wave or traveling-wave accelerator. As described in Chapter 3, microwave power needed to accelerate electrons in a medical linac is generated in an RF source, which is part of the RF system of the RT unit. The RF source can be a magnetron for low- to mid-energy linacs (4 to 10 MeV) or a klystron for high-energy medical linacs (10 to 25 MeV).

Medical linacs differ from other commercial linacs, such as those used in industrial applications, and from those built for high-energy physics experiments, in that they have to satisfy certain requirements to achieve the necessary performance in a radiation oncology treatment unit. Some of these requirements are:

1. *Dose-rate output.* Higher dose-rate helps in shortening the treatment session with the accompanying more convenience for the patient and increase in treatment facility throughput. The dose-rate output de-

pends on the linac design and its efficiency and is limited by the power capability of the RF source.

2. *Output stability.* The consistency of dose-rate output is very essential since it directly affects the amount of radiation dose delivered to the tumor that has to follow accurately the treatment plan for a given patient. This requires a dose-rate that is consistent and stable with time during the treatment and also stable with the gantry rotation. To achieve this stability, the linac cooling water is held and regulated at a nominal temperature (30° to 40°C, depending on the system). Additionally, the output dose-rate is sensed by a dose monitoring component (such as an *ionization chamber*) that is a part of a feedback system that changes the beam parameters, such as its pulse repletion frequency or the pulse width.

3. *Precision of output beam.* The treatment fields (X-rays or electrons) need to match the tumor volume precisely at the 1mm level. In order for the beam shaping techniques to work best, the beam spot at exit from the linac should be as small as possible, preferably in the range of 1 to 2 mm, but not more than 3 mm. The ultimate and ideal source of radiation would be a point source, which is an aim to design for but obviously a theoretical limit.

4. *Size and length of the linac.* To achieve good mobility of the gantry structure, as well as ease in shielding the linac, it is important to optimize the design of the linac and the choice of the operating frequency to accomplish this requirement. For these reasons, as will be pointed later in this chapter, some of the specialized RT machines, such as the robotic-based CyberKnife and the mobile intraoperative Mobetron, use linacs operating at the X-band frequency range (8 to 12 GHz) while the majority of RT machines that use linacs operate in the S-band frequency range (2 to 4 GHz). Although the S-band linacs are larger than the X-band ones, they have the advantage of the availability of S-band RF sources with more power capabilities than the X-band RF sources currently available for such RT machines. A small number of RT machines use linacs operating in the C-band frequency range (4 to 8 GHz), with linac sizes smaller than the S-band and larger than the X-band ones [8].

5. *Cost of manufacturing.* It is a known fact that there is an unsatisfied demand for radiation therapy machines, especially in many of the developing countries. So it is my hope that the up-and-coming new generation of accelerator designers will fulfill this need. They should help cancer patients around the world with new designs of linacs that lend

themselves to more manufacturability and the use of robotic assembly to allow for linac production in large quantities and at a reasonable cost.

5.5 Clinical Use of Linacs in Radiation Therapy

5.5.1 Clinical Requirements

Modern radiation therapy is now one of the most powerful tools for the treatment of localized tumors. However, there are certain clinical requirements that have to be satisfied to achieve successful treatments. The most fundamental requirement is to deliver the necessary dose to kill the cancer cells and at the same time minimize the dose to the surrounding healthy tissue.

Some of the organs adjacent to a tumor can be relatively sensitive to radiation damage, such as the spinal cord, salivary glands, lungs, and the eyes. These must be given special consideration during treatment and in the treatment-planning phase. This is a challenging requirement since, to eradicate tumors within the human body using ionizing radiation, the radiation beams usually traverse normal tissue with the consequent possibility of damaging normal cells and the potential for complications as a result of treatment.

The radiation therapy machine should be capable of delivering a dose that has the following characteristics:

1. The dose should be sufficient to destroy the tumor, to stop the cancerous process, and to prevent the relapse of the disease.
2. The dose should be as uniform as possible over the targeted area to be irradiated.
3. As low a possible dose should be delivered to the healthy tissue surrounding the tumor.
4. The dose rate should be high enough to minimize the time of irradiation and the chances of patient movement.

Other requirements for a radiation therapy machine, in addition to the above, include the stability and repeatability of its dose rate, reliable performance of dose delivery with minimum downtime, the ability to vary the energy of its X-ray or electron treatment beam, and, of course, the patient's safety against any mechanical injury.

Over the last forty years, a great deal of development has occurred and new approaches have been implemented to improve efficacy, safety, and accuracy of dose delivery to the cancer patient. In this chapter, we cover briefly a number of these techniques. Some of them are now common in the every day radiation therapy treatments offered, and some are more specialized.

5.5.2 Treatment Planning and Simulation

The beam delivery step is only one step within the process of radiation therapy. Some of the other essential steps in this process are as follows:

1. Diagnosis and clinical evaluation. This step confirms the existence of the tumor and determines at what stage it is.
2. Modality decision making. The options include surgery, chemotherapy, radiation, or a combination of two or three of these.
3. Obtainment of the necessary images to use in treatment planning. The sources of imaging can be computed tomography (CT), magnetic resonance imaging (MRI), ultrasound, single-photon emission computed tomography (SPECT), or positron emission tomography (PET). With recent advancement in medical image processing, it has become possible to *fuse* images from more than one of the image modalities to form a comprehensive three-dimensional image.
4. Target volume localization. In this crucial step, the tumor boundaries are defined as the targeted volume to be irradiated.
5. Simulation. A machine called the *simulator* is used to emulate the geometry of the treatment unit. It uses diagnostic X-rays to localize the planning target volume and also the critical normal tissues. It also helps determine the optimal placement of the radiation beams on the patient.
6. Radiation treatment planning. In this step, the data obtained from the preceding steps are used to design the treatment fields' geometry, to calculate the dose distribution for each field, and to devise the dose fractionation plan. In many cases the treatment plan includes the use of a variety of beam energies and field sizes. In general, treatment planners attempt to optimize the dose distributions achievable with a given treatment strategy to deliver an effectual dose of radiation to a target volume, while minimizing the amount of radiation absorbed in healthy tissue.

5.5.3 Dose Fractionation

There are multiple approaches followed in radiation therapy aimed at sparing healthy tissues. One of the most effective approaches is *fractionation*. In this protocol, the overall dose is delivered in fractions, or sessions, over a protracted period of time. The dose per fraction is chosen during treatment planning, at an optimal value in order to achieve a high probability of curing the tumor and reasonably low probability of affecting normal tissue. It is clear that the probability of tumor cure increases with increasing radiation dose. However, a careful choice of dose for each fraction takes advantage of the difference in response to irradiation between the tumor tissue and the normal tissue, as shown in Figure 5.5 [9].

Fractionation takes advantage of the differential ability between the tumor cells and the healthy cells for repair. Let us consider two groups of cells, namely, the abnormal group of cells (tumor tissue) and a normal group of cells (healthy tissue). We will assume that they both got the same dose in their daily fractions. After a fraction of the dose is delivered, and during the normal break between radiation sessions (24 to 48 hours), the normal cells in general have a better ability to recover than the abnormal cells. This differential advantage becomes more and more effective with repeated sessions and the breaks between them. Figure 5.6 [10] demonstrates this effect after several successive fractions (e.g., seven fractions). As can be seen in this illustrative example, a large percentage of normal tissue survives while the majority of tumor cells are destroyed when fractionation is employed. Thus, it is clear that in fractionated radiation therapy, if the normal cells receive sublethal radiation, the damage can be repaired more efficiently during prolonged treatment course.

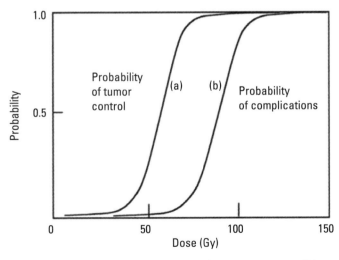

Figure 5.5 Dose-response curves for tumor tissue (a) and normal tissue (b) as a function of dose. (*From:* [9].)

5.5.4 Rotational Therapy

In rotational therapy, the *gantry* rotates through a given arc or a full 360° around the patient who lies on a *treatment couch*. The patient is positioned such that the center of the tumor is located at the *isocenter*, a point defined by the orthogonal intersection of the axis of rotation of the gantry (which is parallel to the patient) and the therapy beam axis (which is normal to the patient).

There are two basic configurations for the RT machine. In one configuration, shown in Figure 5.7(a), the linac is mounted in the gantry perpendicular the patient. This is the simplest configuration for a radiation therapy machine. It eliminates the need for an intricate beam transport system. This is typically used in photon therapy for treatment photon energies between 4 and 6 MV. On the other hand, in higher energy machines, the linacs, which are longer, are located in the gantry parallel to the gantry axis of rotation, as shown in Figure 5.7(b). An intricate beam transport system is then used to transport the electron beam from the accelerator to the treatment head. This configuration is usually used in RT machines delivering both X-ray or electron beams and at multiple energies.

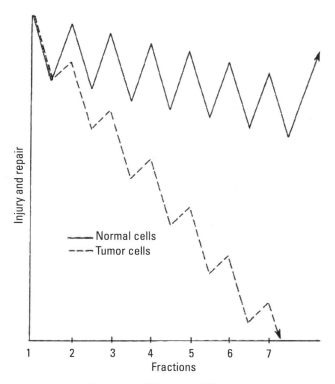

Figure 5.6 Fractionation exploiting the differential ability to repair between the tumor cells and the healthy cells. (*From:* [10].)

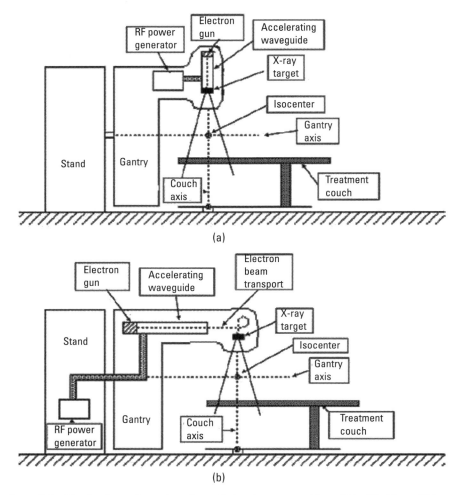

Figure 5.7 Two basic configurations for radiation therapy machines. (*Modified from:* [9].)

5.6 Conformal Radiation Therapy

The goal of *conformal radiation therapy* (CRT) is to deliver a high dose to a volume that closely conforms to the shape of the patient's tumor volume and at the same time minimize the dose to any neighboring sensitive organs. This translates to accurately identifying both the exact shape and location of the tumor. The development of conformal radiation therapy allowed the use of the maximal applicable tumor doses without increasing radiation-induced complications for the patients. This requires good imaging, accurate radiation dose calculation, computer-optimized treatment planning, and computer-controlled delivery of precisely directed treatment beams.

Conformal radiation therapy became possible with some of the recently developed techniques. Tumors usually have an irregular shape, but thanks to those techniques, the radiation oncologist is capable now of shaping the targeted volume exactly to the size and shape of the tumor. Some of the most effective tools and techniques are the multileaf collimators (MLCs), intensity-modulated radiation therapy (IMRT), adaptive radiation therapy (ART), image-guided radiation therapy (IGRT), stereotactic radiosurgery (SRS), and tomotherapy. We will discuss in this section the first three since they go together and complement each other. We will discuss image-guided radiation therapy and tomotherapy in later sections.

5.6.1 Multileaf Collimators (MLCs)

As mentioned before, a major limitation to the efficacy of radiation therapy is the risk of creating undesirable complications by the irradiation of healthy tissue. Conventional radiation therapy machines have been using collimation and shaping techniques to be able to conform to the tumor shape. Downstream from the beam source, sets of high atomic-number metal, such as tungsten, are used to shape the X-ray field into rectangular shapes with different sizes. These collimators (commonly referred to as *jaws*) usually remain stationary during treatment. Additional beam shaping is accomplished through the use of a combination of these jaws and secondary custom beam blocks attached to the treatment head downstream from the collimators.

More recent RT machines have been using computer control to change the position of the jaws from one treatment field to another. The next advancement was to replace the field-shaping beam blocks with the MLCs. They consist of a large number of collimating shielding blocks or *leaves* that can be driven automatically, independent of each other, to generate a field of any shape, conforming the beam to each patient's tumor and, thus, sparing normal tissues. The MLCs have benefited many hospitals offering radiation treatments. They save time and cost compared to the use of specially shaped beam blocks. Additionally, the patient setup time has decreased, allowing greater patient throughput. Similar to the jaws, the leaves are made of high atomic-number metal, such as tungsten. Figure 5.8 shows a schematic simplified representation of the MLC and its position downstream from the two sets of jaws (the upper jaws, or Y-jaws, and the lower jaws, or X-jaws). Only a few leaves are shown and one of the Y-jaws has been omitted for clarity [11].

The larger the number of leaves and the smaller the width of each leaf, the more closely the radiation field at treatment will conform with the contour of the field specified in the treatment plan and the smoother that contour can be. For this reason, the number of leaves has been increasing and their width has been decreasing. Typically MLCs used to have 20 to 80 leaves and a width of 1

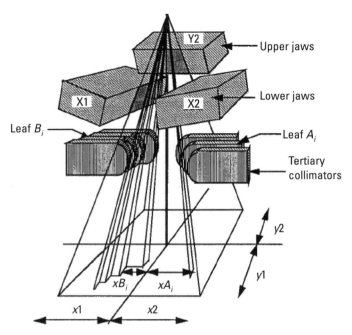

Figure 5.8 A simplified schematic showing the MLC as the third set of collimators. (*From:* [11].)

cm. More advanced MLCs now have 120 to 160 leaves and a width of 5 mm. I list here two examples:

1. One example of an MLC is the Millennium MLC offered by Varian Medical Systems. One of the options has the following characteristics: 120-leaf collimator, field size 40 × 40 cm, and a 5-mm leaf width (projected at isocenter). (See Figure 5.9 [12].)
2. Siemens offers different types of MLC, including the 160 MLC™ Multileaf Collimator. It has 160 leaves, each has the width of 5 mm (projected at isocenter). (See Figure 5.10 [7].)

5.6.2 Intensity-Modulated Radiation Therapy (IMRT)

The computer-controlled MLC discussed above is an essential tool to modulate X-ray beams for IMRT as a means of providing dose distributions that conform to target volumes. Through the use of IMRT, the radiation oncologist is capable of increasing the radiation dose to the areas that need it and reduce radiation exposure to specific sensitive areas of the surrounding normal tissue. MLC are used in conjunction with elaborate computer programs that calculate the re-

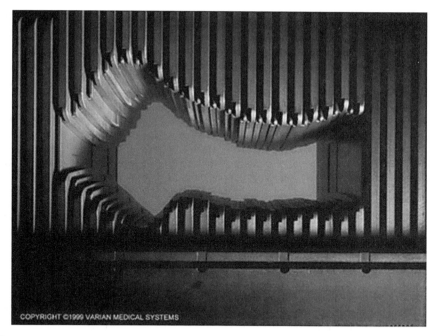

Figure 5.9 120 Millennium MLC™ Collimator offered by Varian Medical Systems. (*From:* [12].)

Figure 5.10 160 MLC™ Multileaf Collimator offered by Siemens. (*From:* [7].)

quired number of beams and angles of the radiation treatment and determine the sequence of leaf positions used to produce the desired modulation. In the IMRT technique, the radiation beam can be looked at as if it is broken up into many *beamlets*, and the intensity of each beamlet can be adjusted individually. The MLCs can be stationary or can move during treatment, allowing the intensity of the radiation beams to change during treatment sessions. The sequence of treatment field segments adds up to the desired dose distribution. This kind of *dose modulation* allows different areas of a tumor or nearby tissues to receive different doses of radiation.

Figure 5.11 shows a simplified example of IMRT. The treatment is comprised of five treatment fields at different angles. Each field is the sum of three segments.

5.6.3 Adaptive Radiation Therapy (ART)

The position of the tumor and its shape can change during a patient's radiation therapy sessions or throughout the course of treatment, which can be a few weeks long. Some of the sources of variations are:

1. Tumor shrinkage;
2. Weight loss or gain;
3. Changes in hollow organ or cavity filling;
4. Respiratory motion of the lung and adjacent organs.

Because of new advancements in RT, adaptive radiation therapy can handle these variations. ART is basically a closed-loop process in which the treatment plan can be modified using a systematic feedback of measurements that confirm the precise position of the patient's tumor during the individual session and over the course of therapy. This requires the accurate determination of the position of the target tissues. This is done by the implantation of what are known as *fiducial markers*, or the use of imaging techniques, such as CT and MV or KV X-ray imaging. Also, it is important to verify the actual delivered

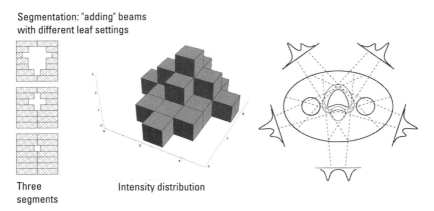

Figure 5.11 A simplified example of IMRT.

dose of radiation in each session. There are multiple RT delivery systems that combine the imaging with radiation. ART thus allows the radiation oncologist to increase the amount of radiation that can be delivered to the tumor while reducing the risk of excess radiation of tissues surrounding or near to the tumor. Two examples of these systems are Accuray's Hi-Art® TomoTherapy system and the CyberKnife® system; both are covered in the following section.

5.7 Image-Guided Radiation Therapy (IGRT)

The success of cancer radiation therapy delivery is founded on maximizing the radiation dose to the tumor while minimizing the effects of radiation on the surrounding tissue. This is most critical when the tumor's surroundings include sensitive tissues and organs such as eyes, spinal cord, or lungs. Achieving this objective is very dependent on the accuracy and precision of dose delivery. One of the recent advancement in RT dose delivery techniques to achieve the improved accuracy and precision is IGRT.

5.7.1 Need for IGRT and Treatment Verification

As mentioned before, radiation therapy for most patients involves a series of RT sessions administered, typically, for five days per week for as long as five or six weeks. During those weeks the patient can gain or lose weight. Also, the accuracy of treatment can be compromised by patient's motion between sessions (such as changes in how the patient is positioned on the treatment couch from one session to the next) or during the treatment session (for example movement due to pain or breathing motion). IGRT is a technique that is used in combination with other methods, including CRT and IMRT to improve the accuracy of treatment and compensate for sources of variation in the tumor position during treatment. IGRT is based upon the concept of incorporating one of the imaging tools into the radiation treatment machine in order to improve the targeting accuracy of radiation therapy and ensure that the patient will be in exactly the same position every day relative to the machine delivering the treatment. Different techniques are used to implement the IGRT efficiently. We cover in this section two of the recently developed examples representing RT machines. These are the TomoTherapy®unit—the Hi-Art® and the CyberKnife®. They are both produced by Accuray and both represent RT machines that are inherently IGRT systems. Before we discuss these recent techniques, let us first review some of the techniques that have been developed over the last few decades to improve imaging and treatment verification.

5.7.2 Portal Films

In the beginning, the only method that was commonly available for verification of the proper positioning of the patient was the use of the megavoltage treatment beam and radiographic film, known as a *portal film*, to produce a radiograph once the patient was positioned on the treatment couch [13]. The primary purpose of port filming is to verify the treatment volume under actual condition of treatment. It provides only a minimal level of detail about the patient's position based on bony landmarks, and cannot visualize tumors within soft tissues. The image quality with the megavoltage X-ray beam (especially for photon energies greater than 6 MV) is poorer than the diagnostic KV X-rays, which offer higher contrast for soft-tissue. Other limitations of portal film include the delay in viewing because of the time required for processing; in many cases, it is actually impractical to do port films for each treatment.

5.7.3 Electronic Portal Imaging Devices

Many different devices have been explored since the early 1980s as alternatives to portal films, and a few were eventually implemented. As a result, these days more and more of the radiographic films are being replaced with *electronic portal imaging devices* (EPIDs), which overcome some of the shortcomings discussed above. With the EPID it is possible to view the portal images instantaneously on a computer screen, providing real time, digital feedback. Daily imaging for treatment localization and verification are becoming more feasible [13]. These images can then be stored for later viewing, distribution, or archiving. EPIDs are typically attached onto the gantry on the opposite side of the isocenter relative to the radiation source.

5.7.3.1 TV Camera-Based EPIDs

One of the common electronic portal imaging approaches is based on the use of a TV camera. In this video-based system, the beam transmitted through the patient excites a metal fluorescent screen. A front-silvered mirror, placed diagonally, reflects the fluorescent light by 90° into the video camera, as shown in Figure 5.12. The images are acquired and digitized, and the data can be further processed to improve the image quality.

5.7.3.2 Flat-Panel EPIDs

One of the recent developments in the EPIDs is the use of an imaging system–based amorphous silicon flat-panel comprised of arrays of photodiodes. Within this unit, a scintillating screen converts the X-ray incident on it to visible light. The light generated by the X-ray converter (scintillator) is detected by the array of photodiodes implanted on the amorphous silicon panel (see Figure 5.13) [14]. The photodiodes integrate the light into charge captures that can be dis-

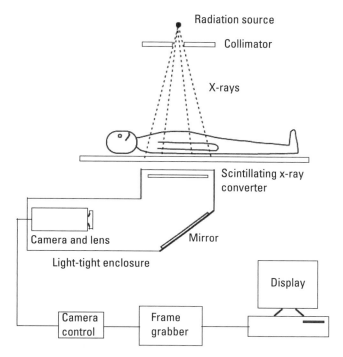

Figure 5.12 Schematic illustration of a camera-based EPID. (*From:* [13].)

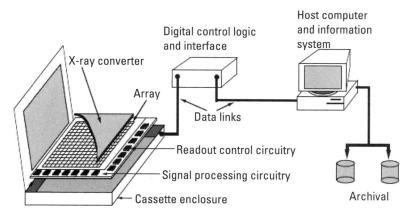

Figure 5.13 Elements of an active matrix flat-panel imager [14].

charged during readout, allowing a current to flow from the photodiodes to the external circuit.

The flat panel-based system has superior image resolution compared to the video-based systems. This helped improve the efficiency and the overall quality IGRT and treatment verification.

5.7.4 Tomotherapy

Tomotherapy a is term derived from *tomography* and *therapy*. Literally, it means slice therapy. TomoTherapy® machines built by Accuray Inc. combine the precision of CT and the capabilities of IMRT into a single radiation therapy unit, known as the Hi-Art® system. This makes it a truly IGRT machine.

Two possible approaches can be used to achieve both the imaging and the therapy beam delivery. The first approach uses a megavoltage X-ray radiation source, a linac, for therapy and a kilovoltage X-ray source for the CT imaging. In the second approach the linac is used as the source of radiation for both the treatment phase as well as the CT imaging.

The first approach is shown schematically, as an artist's conception, in Figure 5.14. The machine consists of a CT-like gantry with a kilovoltage X-ray source and a linac both rotating around the patient. There are two detectors: one for the KV beam and one for the MV beam. The linac's megavoltage x-ray beam is used to deliver the IMRT treatment [15, 16].

The inclusion of an onboard kilovoltage CT expands the possibility of planning, positioning, and verification. However, the use of onboard kilovoltage X-ray source increases the cost and complexity of the unit. For this reason the Hi-Art® shown in Figure 5.15 implements the second approach, namely, using the linac as the source of radiation for both treatment and imaging phases.

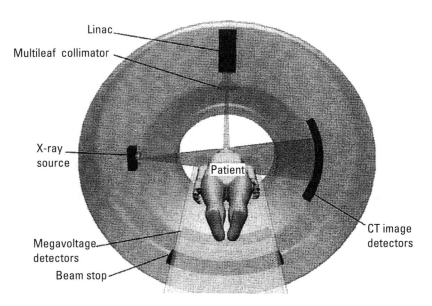

Figure 5.14 Schematic representation of the concept and main components of a tomotherapy machine.

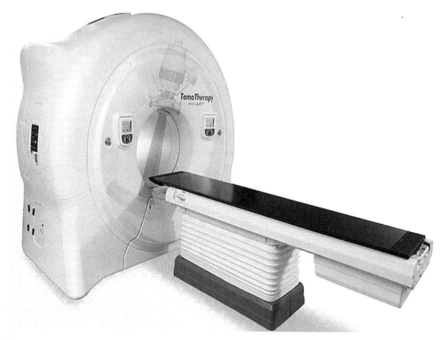

Figure 5.15 Hi-Art® system manufactured by Accuray Inc.

During the treatment the patient is translated in a precise, controlled way, using a couch that moves in synchronism with the gantry rotation to realize *helical delivery*. Figure 5.16 illustrates the concept of helical beam delivery.

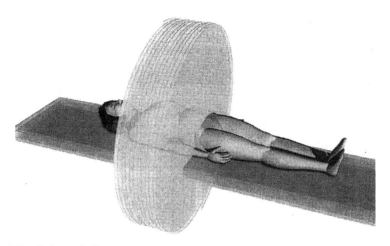

Figure 5.16 Schematic illustration of the concept of helical beam delivery.

TomoTherapy® uses a two-step process that incorporates image-guided targeting of the tumor together with radiation treatment in one seamless progression. The two-step process usually takes about 15 minutes and is as follows:

1. *Image-guided positioning.* Before each treatment, a CT image is acquired on the TomoTherapy® unit. This localizes a patient's anatomy (identifies the tumor size and shape) and verifies the target location against previously stored CT images, resulting in a high accuracy of pretreatment positioning and tumor targeting. If the tumor size or shape changes, then treatment parameters can be modified to adapt to these changes.

2. *Treatment delivery.* After the targeting procedure has verified the target location, the TomoTherapy Hi-Art System® uses a helical 360° rotating beam of radiation, whereby the intensity of the radiation treatment beam is continuously modulated to conform to the shape of the tumor while minimizing radiation exposure to vital structures that may be nearby.

5.7.5 Robotic Radiosurgery: CyberKnife®

The CyberKnife® [16,17] is a *radiosurgery* system designed to treat well-defined tumors. It is most often used to treat malignancies located in the brain, spine, head, and neck. However, more recently, this RT machine has been successfully used to treat tumors at different parts of the body.

The CyberKnife® made by Accuray Inc. [16] uses image guidance and a robot to achieve the precision of delivery, and thus it is particularly useful for tumors that are close to critical structures. The system's X-ray cameras monitor movement during treatment by tracking small markers implanted in the tumor or by tracking the body's skeletal structures. The robotic arm is fitted with a linac and can aim many small radiation beams at the tumor from multiple different angles (see Figure 5.17). This allows the radiation oncologist to give a high radiation dose to the tumor and at the same time maintain a low radiation dose to the surrounding normal tissues. The robotic arm automatically compensates for any movement to ensure accurate delivery of each radiation beam. The linac used can fit in the robotic arm because it operates in the X-band frequency range (8–12 GHz). This makes the linac smaller in length and volume and uses a smaller magnetron compared with most linacs used in radiation therapy, which operate in the S-band frequency range (2–4 GHz).

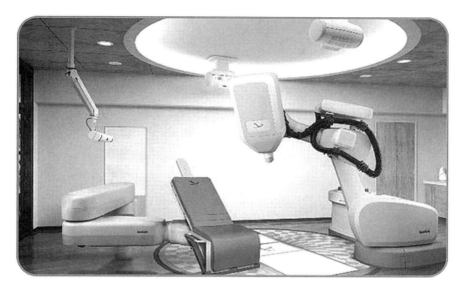

Figure 5.17 Accuray's CyberKnife System®. (Courtesy of Accuray.)

5.8 Stereotactic Radiosurgery

Stereotactic radiosurgery (SRS), also known as *stereotactic radiotherapy* (SRT), is a technique that allows the radiation oncologist to precisely focus beams of radiation to destroy certain types of tumors, both benign and malignant. It delivers high doses of radiation to small tumors with well-defined edges. The SRS delivers radiation therapy in fewer sessions than the conventional RT. It uses smaller radiation fields that can be delivered in just one session or in a few sessions. Because of the high dose delivered in each session, SRS units use extremely accurate image-guided tumor targeting and precision patient-positioning systems. This is done sometimes with the help of rigid immobilizing devices, such as with a head frame in the treatment of brain tumors. Examples of SRS systems are the CyberKnife®, discussed above, and the Gamma Knife ™, which we discuss next.

5.8.1 Gamma Knife ™ Radiosurgery

The Gamma Knife ™, shown in Figure 5.18, is one example of radiosurgery machines [18]. It is typically used for treating brain tumors by accurately focusing many beams of high-intensity gamma radiation to converge on the targeted tumor (see Figure 5.19). Each individual beam is of relatively low intensity, so the radiation has little effect on intervening brain tissue and is concentrated only at the tumor itself. This machine typically contains 201 cobalt 60 sources

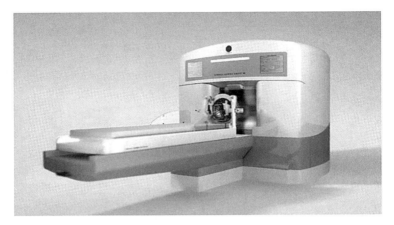

Figure 5.18 Elekta's Gamma Knife ™ [18].

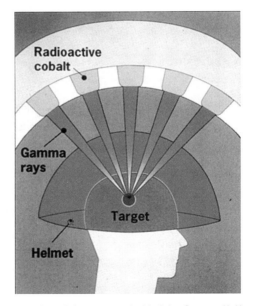

Figure 5.19 Simple illustration of the concept behind the Gamma Knife ™.

placed in a circular array in a heavily shielded assembly. The patient wears a special helmet that is surgically fixed to the skull, so that the brain tumor remains stationary at the target point of the gamma rays. Thus, the tumor receives a substantial dose of radiation, while surrounding brain tissues are relatively spared.

5.9 Intraoperative Radiation Therapy

Intraoperative radiation therapy (IORT) is the use of radiation therapy during a surgical procedure. In this special RT technique, a radiation dose is delivered in a single session to a surgically exposed internal tumor bed after the removal of the tumor by a surgeon. Thus it allows the delivery of high radiation doses, which can be as high as 10 Gy, to a tumor while sparing the nearby normal tissues. During the surgical procedure, the surgeon shows the radiation oncologist the areas of the tumor bed where tiny fragments of tumor may remain. After the surgeon removes most or all of the visible parts of a tumor, there is always the chance that microscopic cells are left behind in the tumor bed. These cells would most likely repopulate the tumor mass and eventually causing recurrence of cancer. Before the delivery of radiation, the sensitive normal tissues are either shielded or displaced out of the radiation field, and then the radiation oncologist directs a single high dose of radiation to the area most at risk for residual disease.

Because the target area is exposed, electron beam radiation is more suitable for IORT than X-ray radiation. IntraOp Medical Corporation (Santa Clara, CA) has developed a dedicated IORT machine, the Mobetron [19]. The Mobetron is a mobile unit dedicated for IORT that produces only electron beams and can be moved among several operating rooms. Figure 5.20, shows the Mobetron.

The accelerator system in the Mobetron consists of two collinear linacs. Each linac is of the standing wave-type operating at 9.3 GHz. The use of X-band technology in the Mobetron [20] makes its linacs significantly lighter and smaller than S-band linacs used in most of the conventional RT machines. The Mobetron treatment head houses the X-band linacs, the magnetron, the modulator, RF loads, the circulator, and other RF circuitry. The beam stopper shown in Figure 5.20, above, is an integral part of the Mobetron and is always positioned to intercept the bremsstrahlung radiation generated in the forward direction of the electron beam. The unit can produce electron beams with four nominal energies of 4, 6, 9, and 12 MeV at two dose rate settings (250 cGy/min and 1,000 cGy/min).

The small size of the X-band SW linac used in the Mobetron allows for lead shielding that is placed inside the treatment head. However, as an additional precaution to avoid the exposure of the surgical and oncology staff to any radiation leakage, the staff leaves the surgical suit for the few minutes during the application of the dose. The control of the dose is done through a controlling unit placed outside the operating room while monitoring the patient through a glass window and TV cameras.

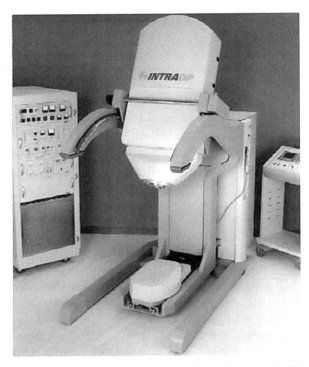

Figure 5.20 The Mobetron, an IORT unit by IntraOp Medical Corporation (Santa Clara, CA).

5.10 Concluding Remarks

The RT linac application covered in this chapter represents one of the classic examples of how a device that was initially developed for basic research, the RF linac, had made its way to hospitals and oncology centers to offer a significant contribution in the fight against cancer. As was presented in this chapter, in addition to the availability of the linac, it took a succession of new ideas and new technologies especially in the field of medical imaging to develop the current radiation therapy machines. These advanced RT units are now capable of selectively targeting tumors with great precision in three dimensions or even four dimensions (with time/motion being the fourth dimension).

A growing number of research centers are working on developing the up-and-coming field of particle therapy, which uses protons or carbon ions instead of X-rays to target cancerous tumors. The chief advantage of particle therapy is the ability to more precisely localize the radiation dosage when compared with the conventional RT using X-rays. We discuss particle therapy in Chapter 6.

References

[1] Strydom, W., et al. "Electron beams: Physical and Clinical Aspects," *Radiation Oncology Physics: A Handbook for Teachers and Students*, E. B. Podgorsak (ed.), Vienna, Austria: International Atomic Energy Agency (IAEA), 2005, pp. 273–299.

[2] Baltas, D., et al., *The Physics of Modern Brachytherapy for Oncology*, New York: Taylor & Francis, 2006.

[3] Devlin, P. M. (ed.), *Brachytherapy: Applications and Techniques*, Philadelphia: Lippincott Williams & Wilkins, 2006.

[4] Thomadsen, B. R., et al. "Brachytherapy," AAPM Medical Physics Monograph #31, Madison, WI: Medical Physics Publishing, 2005.

[5] Glasgow, G. P., "Cobalt-60 Technology," in J. Van Dyk (ed.), *The Modern Technology of Radiation Oncology*, Madison, WI: Medical Physics Publishing,1999, p. 313.

[6] Khan, F. M., *The Physics of Radiation Therapy*, 2nd ed., Baltimore: Williams and Wilkins, 1994, p. 418.

[7] Siemens Medical Inc. company website, www.medical.siemens.com/.

[8] Ishkhanov, B.S., et al., "Conceptual Design of the Miniature Electron Accelerator Dedicated for IORT," *Proceedings of RuPAC XIX*, 2004, Dubna, Russia, pp. 474–476.

[9] Podgorsak, E. B. (ed.), *Radiation Oncology Physics: A Handbook for Teachers and Students*, Vienna: International Atomic Energy Agency (IAEA), 2005.

[10] Kligerman, M. M., "Principles of Radiation Therapy," *Cancer Medicine*, J. F. Holand and E. Frei (eds.), Philadelphia: Lea and Febiger Publisher, 1973, pp. 541–565.

[11] Boyer, A. L., et al., "Beam Shaping and Intensity Modulation," in J. Van Dyk, ed., *The Modern Technology of Radiation Oncology*, Madison, WI: Medical Physics Publishing, 1999, p. 437.

[12] Varian Medical, www.varian.com/.

[13] Reinstein, L. E., et al., "Radiotherapy Portal Imaging Quality," AAPM Report of Task Group No. 25, New York: American Institute of Physics, 1988.

[14] Antonuk, L. E., "Electronic Portal Imaging Devices: A Review and Historical Perspective of Contemporary Technologies and Research," *Phys. Med. Biol.* Vol. 47, 2002, pp 31–65.

[15] Olivera, G. H., et al., "Tomotherapy," in *The Modern Technology of Radiation Oncology*, Madison, WI: Medical Physics Publishing, 1999, p. 521.

[16] Accuray Inc., www.accuray.com/

[17] Schweikard, A., H. Shiomi, and J. Adler, "Respiration Tracking in Radiosurgery," *Medical Physics*, Vol. 31, No. 10, 2004, pp. 2738–2741.

[18] Elekta, A. B., www.elekta.com/.

[19] IntraOp Mobetron, www.intraopmedical.com/.

[20] Hanna, S. M., "Applications of X-Band Technology in Medical Accelerators," *Proc. 1999 Particle Accelerator Conference*, New York: IEEE, 1999, pp. 2516–2518.

6

Accelerator-Based Industrial Applications

It is a fact that the original motivation for scientists and engineers to build high-energy accelerators was to probe deeper into the structure of matter. However, eventually accelerators became the enabling technology for many industrial applications. Thanks to technology (and also people) transfer from research centers and national laboratories to industry in several countries around world, we now have multiple established industries and commercial markets based on the use of accelerators. In this chapter, I try to give you, the reader, a glimpse at some of uses of electron beams and X-rays in manufacturing of many of our everyday-life products.

6.1 Use of Accelerators in Material Processing

The largest industrial use of electron-beam (EB) accelerators is for material processing. Radiation has been used for more than 50 years to improve both bulk and surface properties of polymer resins. Originally the radiation-processing source had been based on a radioactive source, such as cobalt 60. However, electron-beam accelerators offer much higher dose rate and are generally a more cost effective way of delivering dose to the product. The electron-beam linac lends itself to more efficient movement of the product in front of the radiation source and offers more flexibility. Additionally, EB units do not face the security, transportation, and disposal issues that confront the use of radioactive isotopes. It is estimated that there are more than 1,400 high-current EB units in commercial use around the world providing an estimated added value to numerous products of more than $85 billion US [1]. This industrial application is based upon the use of accelerated electron beams for the modification of polymers by *cross-linking*, which is basically joining one polymer chain to another to form three-dimensional chemical links between adjacent polymer segments. Cross-linking renders materials with enhanced properties and characteristics better than the original non-cross-linked materials.

One example of EB cross-linking end-use applications is the implementation of the electron beam in the curing of rubber. A second example, which makes up a large share of the electron-beam market, is the cross-linking of heat-shrinkable films, most widely used in food packaging. Such films extend the shelf life of meat, produce, poultry and dairy products, and provide tamper-resistant packaging. Other examples of the utilization of high-current industrial EB accelerators include surface curing, polymeric precursors used in inks, coatings, and adhesives, insulation on electrical wires, and heat-shrinkable tubing for protecting wire and cable connections, making these products more flame-retardant.

6.1.1 Electron Beam Current and Energy Requirements

In material processing applications of industrial electron beams, we need to design and control two fundamental properties of the accelerated electron beam. The first requirement is concerned with the accelerator beam current. High beam current is desired in industrial linacs because it permits the irradiated product to be exposed to the required dose in a shorter time. Thus, the moving conveyer (carrying the product being processed) can be moved at faster speeds transverse the beam and hence achieves more product throughput. The second requirement is the beam's energy needed to match the type of industrial application. Electrons' penetration into materials are limited by their energy and by the density of the material being processed. Figure 6.1 shows the distribution of the absorbed dose (the energy deposited) for beam energies between 1.0 and 5.0 MeV at different depths (in centimeters of water) as derived from computer simulations [2]. In order to be able to use this curve for different materials, the absorbed dose is described as the *normalized energy deposition*. It is the energy deposited, divided by the *normalized depth*. Note that the normalized depth is the depth multiplied by density and thus its unit then is cm Xg/cm^3 = g/cm^2. Thus the normalized energy deposited is measured in MeV over the normalized depth, and its unit is then MeV cm^2/g.

From the graph in Figure 6.1, we can see that the higher the beam energy, the deeper the penetration. Thus for different EB processing application, the manufacturer should choose the accelerator with the appropriate beam energy. Table 6.1 [1] lists some of the EB applications, their typical beam penetration requirements, and the corresponding electron energy range to realize such depths of deposition. An example of a 10 MeV EB industrial linac used in

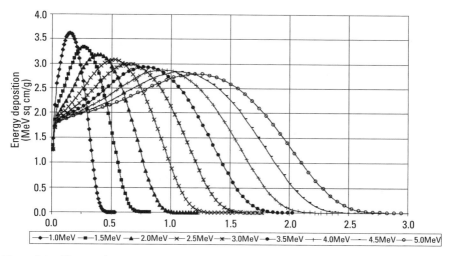

Figure 6.1 Electron beam energy deposition in centimeters of water (1 to 5 MeV) [2].

Table 6.1
Some EB Applications, Corresponding Penetrations, and Required Beam Energies

EB Application	Typical Penetration	Electron Energy
Surface Curing	0.4 mm	80–300 keV
Shrink Film	2 mm	300–800 keV
Wire & Cable	5 mm	0.4–3 MeV
Sterilization	38 mm	3–10 MeV

processing (produced by MEVEX, in Canada) is shown in Figure 6.2 [3]. Note the beam fan-out provision (the scan horn) used to create a "shower" or "curtain" of directed electron beam, thus covering relatively large areas of the material to be irradiated.

It is to be noted that electron beams cross in air for processing and suffer some scattering from colliding with air molecules. However, these effects are normally taken into consideration in designing the EB processing system, including the dose rate used and the distance between the scan horn and the conveyer carrying the product to be processed.

6.1.2 Main Applications for Polymer EB Cross-Linking

6.1.2.1 Surface Curing

Electron-beam processing is an effective surface curing technique especially when dealing with high-volume throughput on a continuous basis. Electrons in

Figure 6.2 A 10 MeV industrial linac from Mevex.

EB curing affect chemicals that are added to cure inks and coatings, often more quickly and less costly than by other methods.

In EB curing surface coatings, the electron beam penetrates down to the substrate irrespective of pigment color. The energy of the electron beam determines its depth of penetration into the product. For curing of thin coatings, a low-energy beam (less than 1 MeV) is satisfactory. However, for thicker products, including bulk polymers in open trays or on continuous belts, a medium-energy (1 to 5 MeV) or a high-energy beam (5 to 10 MeV) would be required.

Both the ultraviolet (UV) and EB curing processes are types of radiation that have been used extensively in surface curing. They both provide the energy needed to polymerize and solidify the dilutant in the coating material. This reactive dilutant is used to provide the viscosity needed to easily apply the coating. EB surface curing has certain advantages over other alternative manufacturing processes, as follows:

1. In the EB processing, the energy is delivered directly to the polymers being cured not through heat applied externally. Energy emitted from an EB unit is directly absorbed in the coating and causes the chemical changes that convert a liquid coating into a dried and cured material. Thus it is an efficient process for coating operations that helps the manufacturing facilities in reducing their energy consumption. The energy efficiency of EB processing can be contrasted with conventional uses of solvent drying or water-based technologies. When comparing the energy demand needed to dry a water-borne coating, EB curing would require two orders of magnitude less energy to attain comparable properties.

2. In the EB surface curing technologies, the products do not lose any mass through evaporation and thus avoid volatile organic compounds

(VOCs) that can be created as a result of other alternative manufacturing processes. This feature is particularly important with the increasing awareness by manufacturers of the effects of greenhouse gas emissions and their efforts to reduce air pollution.

3. Because electron beam processing is not a thermal means of energy transfer and takes place at near-ambient temperatures, EB drying of inks can be used on heat-sensitive substrates, such as plastic films, minimizing concerns over film distortion. The EB technique is also used on a variety of substrates with low heat requirements necessary to achieve a certain finish.

4. Some of the above useful features are shared with UV curing. However, an advantage to EB surface curing and cross-linking of coatings over the use of ultraviolet radiation is that electrons have the ability to penetrate pigments, whereas UV does not.

6.1.2.2 Wire and Cable Insulation and Heat-Shrinkable Films

One of the applications of the electron beam cross-linking is in the manufacturing of the insulation jacketing on wires and cables. Cross-linking prevents insulation from dripping off an over-heated wire, as could result from a short circuit, or when exposed to the high heat of an automotive engine or even a fire. Normally, there are specialized fixtures to transport the wire under the electron beam using multiple passes. The entire process can run at several hundred meters per minute [4].

EB cross-linking of heat-shrinkable films makes up a large share of the electron-beam market. Such films extend the shelf life of meat, produce, poultry and dairy products, and provide tamper-resistant packaging. Some of the advantages of EB treatment of flexible packaging are:

- No Volatile Organic Compounds (VOCs) or emissions;
- Instant drying at high production speeds;
- High chemical resistance;
- High abrasion resistance;
- High gloss;
- No softening after curing;
- Low odor and off-taste.

6.1.2.3 Tire Rubber Treatment

The tire industry uses electron-beam to treat its rubber. The rubber is extruded and then irradiated to produce chemical cross-links between polymer chains, bringing them to a gel state. In that gelled state, the elastomers (elastic polymers) gain more elasticity as the chains stretch under stress and retract on release of the stress. This makes the rubber tougher than the noncured rubber and prevents tire cord distortion or strike-through during subsequent molding operations [5]. The ability to tightly control electron beam exposure enables tire manufacturers to only partially cure or cross-link elastomers to the degree desired.

6.1.2.4 Gemstone Irradiation

Electron beam irradiation is used to alter the color of some gemstones so as to enhance their commercial value. An example of frequently irradiated gemstones is topaz. Its linac-irradiated material becomes an attractive "sky blue" after subsequent heat treatment of the greenish or brownish product. Other gemstones that are irradiated for similar effect are: rubellite, quartz, citrine, amethysts, and even diamonds. Since very high doses are often required, care is taken not to overheat the gems while they are being irradiated. High temperatures can cause undesirable changes in the color of the stones, while thermal shock will crack or shatter them. For this reason, the gem materials are water-cooled to prevent elevated temperatures and thermal shocks. Also, some gems must be set aside for weeks or even months to allow for the decay of any induced radioactivity, as small as it may be [6].

6.1.3 Industrial Material Processing Using X-Ray Radiation

In addition to providing electron beams for material processing, accelerators are also used as sources of X-rays for industrial processing of materials. The availability of high-current accelerators made it possible to produce X-rays with dose-rates compatible with industrial applications. The dose-rate for such applications is more than 3 orders of magnitude larger than those used for the cancer radiation therapy discussed in Chapter 5. X-rays, as with gamma rays, can penetrate an order of magnitude deeper than the typical penetration of commercial electron-beam sources. Since these commercial X-ray sources are generated by accelerators, they share the advantage of EB processing units over the gamma-ray sources, as we discussed before. We show in Figure 6.3 a comparison of typical penetrations between gamma rays from a cobalt 60 source, an EB unit (10 MeV accelerator), and an X-ray unit (7.5 MeV accelerator) [7]. Note that the depth here is the normalized depth (density X depth), and thus its unit is g/cm^2. Similar to the industrial EB units, the commercial accelerators producing X-rays for material processing are normally designed for acceleration energies below 10 MeV to avoid inducing radioactivity in the materials being processed.

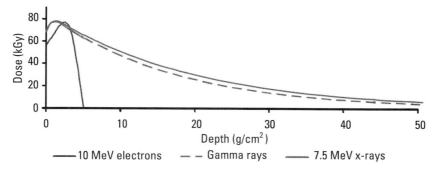

Figure 6.3 Comparing material penetration for EB, gamma rays, and X-rays.

The greater depth of penetration of X-rays allows for treatment of products with large volumes, such as pallet loads of packaged medical devices and food. Since the X-ray irradiation is not accompanied by a significant rise of temperature of the product being sterilized, it eliminates adverse effects on materials due to heat. Thus it is suitable for temperature-sensitive products, such as frozen food.

6.2 Sterilization of Medical Products and Food Irradiation

As early as the end of the nineteenth century and the beginning of the twentieth century, it was realized that the "newly discovered" X-ray radiation (discovered by W. C. Röntgen in 1895) would kill microorganisms. However, the use of this fact had to wait for more than a half century before it started to find commercial application in sterilization and disinfection. Nowadays, electron beams and X-rays derived from electron-beam systems are used in the sterilization of medical products as well as preservation of food products in many countries around the world.

6.2.1 Medical Product Sterilization

Conventional sterilization techniques for medical equipment have mainly used gases, such as the ethylene oxide (Et O), high-pressure steam (autoclave sterilization), or gamma rays. Ethylene oxide gas leaves toxic residue on sterilized items, so a lengthy purging process usually follows it. The autoclave sterilization is unsuitable for heat-sensitive objects. Accelerators offer effective alternative sterilization techniques. Either a beam of electrons or X-rays can be used, but EB sterilization is used more commonly than the X-rays. It is a safe and reliable method for sterilization, especially for disposable medical products. It offers a cost-effective and high throughput alternative by sterilizing medical products in their final packaging with no hazardous chemicals used in the process. It has the fundamental advantage over the gamma sources in that the EB unit can be turned on or off, as required with no need for radioactive waste disposal issues. Accelerators used for EB sterilization produce electron beams with energies up to 10 MeV. Many of the packaged devices have a low bulk density so that the penetration of electrons is sufficient. If needed, packages can be irradiated from two sides, thereby increasing the beam penetration to more than double the single-beam penetration by taking advantage of the overlap of the tail ends of the two depth-dose profiles, as shown in Figure 6.4. To attain more uniform dose distributions in large pallet loads of packaged products, a turntable can be used to rotate the pallet in front of the EB accelerator scan horn.

Figure 6.5 shows a 10 MeV electron beam accelerator produced by L-3 [8] and used for sterilization of medical products. Such a linac would be installed in a special facility shielded with thick concrete walls, as shown in the computer-generated model in Figure 6.6. The shielding requirements and specifications of such facilities are similar to those discussed in Chapter 3 of this book in the

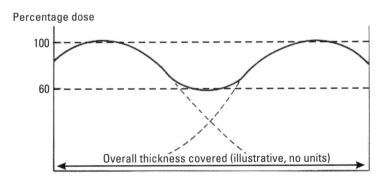

Figure 6.4 Benefit of applying irradiation from two opposite sides.

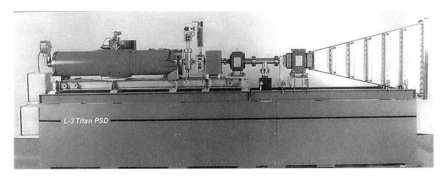

Figure 6.5 An electron-beam accelerator for sterilization of medical products by L-3 [8].

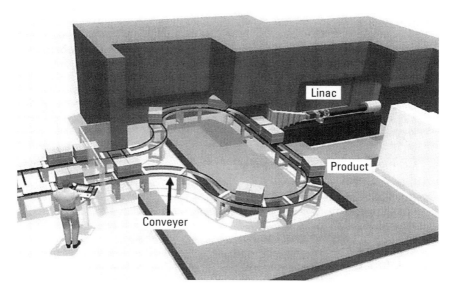

Figure 6.6 A model for a sterilization facility for medical products by L-3 [8].

section that covers linac high-power testing. It is to be noted that the shielding walls have been cut away in the figure to show the path of the conveyer.

6.2.2 Electron Beam Food Processing by Irradiation

Similar to the sterilization of medical products described above, the electron beams and X-rays generated by linac systems are used in the preservation of food products. This ionizing radiation is capable of eliminating food-borne pathogens such as *Escherichia coli (E. coli), Salmonella,* and *Listeria,* from meats, poultry, and other food products, and can also disinfect grains, spices, and animal feed.

Although there are several food decontamination methods, irradiation is considered a high throughput, environmentally clean, and energy-efficient process. One unique feature of radiation decontamination is that it can also be performed when the food is in a frozen state.

6.2.2.1 Required Doses for Food Irradiation

It is important to ensure that food is exposed to enough doses to achieve the required results, and at the same time, that the dose is within the permissible limits. For this reason the dose should be measured precisely and routinely. The absorbed dose in a product depends upon the intensity of radiation, the distance from source, speed of conveyer, and the characteristics of the product being irradiated. Table 6.2 shows different food irradiation applications and the

Table 6.2
Different Food Irradiation Applications and the Doses Used

Low-Dose Applications 0.03 to 1 kGy	
Application	**Dose**
Sprout inhibition in bulbs and tubers	0.03–0.15 kGy
Delay in fruit ripening	0.25–0.75 kGy
Insect disinfestations including quarantine treatment and elimination of foodborne parasites	0.07–1.00 kGy
Medium-Dose Applications 1 kGy to 10 kGy	
Application	**Dose**
Reduction of spoilage microbes to prolong shelf life of meat, poultry, and seafood under refrigeration	1.5–3 kGy
Reduction of pathogenic microbes in fresh and frozen meat, poultry, and seafood	3–7 kGy
Reducing the number of microorganisms in spices to improve hygienic quality	3–7 kGy
High-Dose Applications 10 kGy to 70 kGy	
Application	**Dose**
Sterilization of packaged meat, poultry, and their products that are shelf stable without refrigeration	25–70 kGy
Sterilization of hospital diets	25–70 kGy

doses used [9]. In order to achieve some of these doses, it is required sometimes to use two sources from both sides of the conveyer carrying the food containers, as explained in Section 6.2.1 and as shown in Figure 6.4.

6.2.2.2 Acceptance of Irradiated Foods

It is an unfortunate fact that the pathogens found in the food supply chain, such as *Listeria, Salmonella,* and *E. coli,* have been causing illnesses and deaths and have resulted in major recalls of foods by producers. It is also a fact that today we have available to us diverse and powerful food processing technologies, such as food irradiation, that can help prevent food-borne illnesses. However, there are still some regulatory roadblocks to food irradiation in several countries, in addition to the controversial stigma of radiation issues that somewhat affect consumer acceptance and, hence, the growth in the EB irradiation market. It is hoped that further public awareness will overcome these barriers in the near future.

6.2.3 Mechanism of Killing Pathogens

The effectiveness of EB or X-rays in sterilization and food decontamination processes stems from the role of ionizing radiation in killing harmful microorganisms by breaking the double strand of the DNA (deoxyribonucleic acid) [10]. DNA's double helix encompasses long chain molecules that are also held together by hydrogen bonding as shown in Figure 6.7. The molecular architecture of DNA is self-replicating and, thus, forms the basis of life and genetic encoding. The ionizing radiation generates both free radicals and, in the presence of water, a constituent of all living matter, highly reactive short-lived hydroxyl radicals. Both can lead to the breaking of single- and double-strand, as well as intra-strand cross-links, thus damaging the DNA. With but a few double-strand breaks, the DNA would be no longer able to repair or replicate itself. Thus ionizing radiation can be effective in eliminating pathogens, such as a bacterium or fungus, that can contaminate medical and food products.

6.3 Environmental Applications of Accelerators

Some of the emerging applications for radiation processing can benefit the environment by reducing water and gas pollutants. Radiation processing can remediate polluted waters wastewater in a nonchemical safe way. Flue gas treatment based on electron beam technology can help to mitigate environmental degradation.

6.3.1 Wastewater Treatment

Rapid population growth and increased industrial development have led to the generation of large quantities of polluted industrial and municipal wastewaters. In addition to the conventional wastewater treatment techniques, researchers in several countries around the world have been recently developing techniques

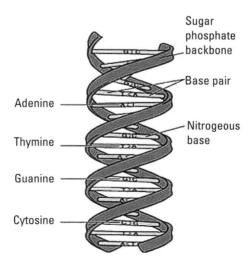

Figure 6.7 DNA double helix.

that employ ionizing radiation for wastewater and industrial waste treatment, reduction of sewage sludge sanitary contamination, and solid agriculture waste transformation [11–13]. Radiation treatment techniques are capable of degrading water pollutants at a rate faster than conventional treatment techniques. The ionizing radiation can be produced by the use of a gamma radiation source, such as cobalt 60 (^{60}Co) or Cesium 137 (^{137}Cs), or the use of an accelerator generating a high-energy electron beam. Because of advantage mentioned in other sections of this chapter, the EB-based units are at the present time used more than the radioactive isotope gamma sources.

Water is not normally reactive, but when it is subjected to ionizing radiation, it produces highly reactive species. As the high-energy electrons impact flowing polluted water, the electrons slow down, lose energy, and react with the water to produce three reactive species (hydrated electrons, hydroxyl radicals, and hydrogen atoms) that are responsible for organic compounds' destruction. These short-lived radicals drive both oxidation and reduction reactions at the same time. High-energy electron beam irradiation is the only process that is capable of forming both highly oxidizing and highly reducing reactive species in aqueous solutions at the same time and in approximately the same concentrations. Furthermore, no other advanced oxidation process has the capability of generating as high an overall free-radical yield per unit of energy input and in a controlled way as high-energy electron beam treatment.

To insure the effectiveness of the EB treatment, the delivery system at the EB-wastewater interaction zone should provide for uniform dose distribution as well as delivery of large amounts of wastewater for EB exposure. Three different schemes for wastewater delivery to the interaction zone are shown in Figure 6.8 [13].

We can see that radiation water processing offers the following advantages:

1. Strong reducing and oxidizing agents;
2. Process controllability;
3. Compatibility with conventional methods.

Since radiation treatment technology represents a viable solution to the problem of wastewater treatment, several countries have already taken the initiative of implementing electron beam in the irradiation of wastewater. Examples are in the Russian Federation [14], the Republic of Korea [15], the United States [16], and Brazil [17]. It is expected that the spread of use of this technology will provide the impetus for the accelerator industry to produce linacs specially designed for this application. It is hoped that the increase in number of units produced would help the cost of such technology to decline to a level that makes it affordable and economically attractive to many countries.

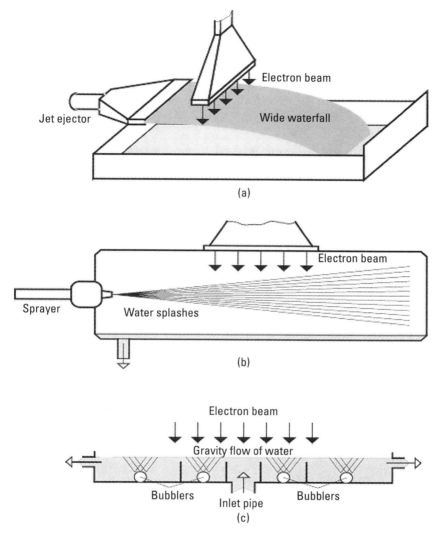

Figure 6.8 Different schemes for EB-wastewater interaction.

This can have a large impact in countries in which water is in short supply and which can benefit from the reuse of treated wastewater for irrigation.

6.3.2 EB Treatment of Flue Gases—Reduction of Acid Rain

Power plants that are using coal and oil to generate electricity and run factories around the world are also producing acid rain by emitting sulfur and nitrogen oxides. These gases are converted to sulfuric and nitric acids in the atmosphere by reactions with water vapor activated by ultraviolet radiation from the sun.

There are several conventional technologies for treating flue gases, which are aimed at controlling the emission of sulfur dioxide (SO_2) and nitrous oxides (NO_x) into the environment. Some of these technologies are the wet, dry, and semi-dry flue gas desulphurization (FGD) and selective catalytic reduction (SCR). These conventional gas-cleaning technologies are complex chemical processes that result in the generation of wastewater and undesired byproducts. These drawbacks can be avoided with the use of the EB technology to treat flue gases [13, 18–21].

The EB flue gas processing is based on the formation of acid vapors under controlled conditions within the power plant. These acidic gases can then be neutralized by injecting ammonia vapor to produce fine particles of ammonium sulfate (($NH_4)_2 SO_4$) and ammonium nitrate ($NH_4 NO_3$), which can be removed from the gas stream by conventional collectors such as electrostatic precipitators (ESPs) or filters. A simplified diagram of EB flue gas processing is shown in Figure 6.9 [21]. The proof of principle of EB treatment of flue gases was demonstrated by Japanese scientists in the 1970s and was further developed to pilot and large scale facilities by researchers in Germany, Japan, Poland, Bulgaria, China, and the United States [18].

The flue gas treatment using EB accelerators has several advantages over currently used conventional methods [21]:

1. It simultaneously removes sulfur dioxide and nitrous oxides efficiently;
2. It is a dry process that is easily controlled and has excellent load-following capability;
3. The pollutants can be converted into useful agricultural fertilizers;
4. The process has relatively low operating cost requirements.

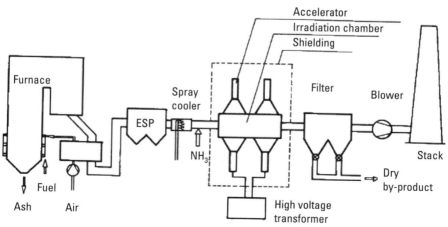

Figure 6.9 A simplified diagram of the EB flue gas processing [20].

6.4 Nondestructive Testing

The goal of nondestructive testing (NDT) is to determine the absence or presence of conditions or discontinuities that may have an effect on the usefulness or serviceability of the object under test without changing or altering that object. Some of the NDT techniques include the use of X-ray radiographic, ultrasonic, eddy current, and thermal infrared testing [22]. NDT can be an important part of quality control (QC) or quality assurance (QA) processes in the manufacturing of critical components. One example where the NDT radiographic techniques are used in the manufacturing of critical components is the detection of flaws that could cause a malfunction in solid-propellant rocket motors.

Historically, early radiographics used gamma rays from radioactive isotopes such as cobalt 60 (^{60}Co), cesium 137 (^{137}Cs), or iridium 192 (^{92}Ir). Because of recent developments in reliable and compact linacs, linac-based NDT devices are replacing many of NDT devices based on radioactive isotopes [23, 24]. NDT linacs (typically in the range of 1 MeV to 15 MeV) have the advantage of the ability to produce megavoltage X-rays. Thus, they can be used in the inspection of thick sections of materials that cannot be inspected using standard X-ray tubes, which are only capable of producing kilovoltage X-rays. Linac also can produce higher dose-rates resulting in improved image capability and shorter exposure times, thus increasing throughput and productivity. That is in addition to eliminating the risk and cost associated with the storage and the disposal of the spent radioactive material, as discussed before in other applications. Most of the NDT linacs operate in the S-band frequency range (around 3 GHz). However, the development of higher frequency linacs operating in the X-band (around 9 GHz) with their small sizes made the portable radiographic system a reality [25–27]. The portable linac systems are used in NDT field inspection of pipelines, ships, bridges, and other critical civil infrastructures. They can also easily be moved to inspect large sections of materials in factories.

In linacs used in the nondestructive testing, the high energy electrons impinge on a metallic target to be converted into X-rays by the bremsstrahlung process (as described Chapter 2, Section 2.5). The target is commonly made of a material having a high atomic number, such as tungsten, tantalum, or gold. The conversion efficiency of electron to X-ray is relatively low and the majority of the electron beam energy (more than 90 percent) is converted into heat. For this reason the target has to be able to tolerate the heat and to dissipate it fast. Consequently, the target is normally water-cooled, and some of the NDT linacs use a layered target comprised of either tungsten or gold and brazed to copper for better heat transfer. One important parameter in the design of the NDT linacs is the spot size of the X-ray emerging from the target. Ideally this should be as close as possible to a "point source." However, there are design limitations on minimum spot size due to scattering of electrons in the target material. The thinner the target is, the smaller the spot size. So using "thin" target is recommended for NDT applications. However, it is important to ensure that any "unconverted" electrons that pass though the target are trapped in some manner, either by a magnetic field or in a low-density material such as carbon. Another desirable feature in the X-ray beam is to have a uniform X-ray field projection on the object. To achieve this, some NDT units use flattening filters similar to the X-ray radiation therapy machines.

Some of the NDT applications require multiple energies for complete inspection of a single part. Hence, a single machine with linac capable of producing multiple energies is desirable for economic radiographic inspection of a wide variety of industrial products. There are several techniques that can be used to accomplish energy variations [28]. These include employing a mechanical switch [29, 30] or an electronic switch [31] built in the linac. However, these add to the linac's complexity and manufacturing costs. The following are two alternative ways of changing the linac energy:

> *Changing the RF power.* As was explained in Chapter 2, the energy of the output X-ray in a linac system is dependent upon the strength of the accelerating field in the linac used. The strength of this field is a function of the RF power fed into the linac. The RF power can be changed by adjusting the operating point of the RF source (a klystron or a magnetron) driving the linac. It also can be adjusted through a variable mismatch connected to waveguide feeding the RF to the linac. A combination of both typically works best since there is a limit as to how much you can adjust the energy using either technique alone.
>
> *Beam loading.* An alternative way to change the output energy of a linac is to change the electron-beam current. The beam current has a loading effect on the linac. When the beam current is increased, the fields in the linac have to do more work to accelerate the additional charge carriers, and this loading effectively reduces the net field in the linac. Thus, the output energy of the accelerated electrons is reduced.

X-ray NDT radiography has benefited from recent developments in medical imaging. One example is the incorporation of the concept of computed tomography (CT). There are two approaches for the implantation of CT scanning in NDT units. In one approach, the linac as the X-ray source and the detector (both on opposite sides of the inspected object) rotate around the inspected object. In the alternative approach, both the X-ray source and the detector are stationary while the inspected object is being rotated and translated between the source and the detector. Many of the new advancements in NDT are being shared with the use of linacs for cargo inspection, and vice versa, as discussed in the next section.

6.5 Security and Inspection Applications

Many ports around the world rely on X-ray scanners to inspect cargo for nuclear materials or weapons and to prevent contraband from entering their countries. Conventionally these have been gamma-ray scanners based on radioactive isotopes such as cobalt 60 or the common dc X-ray tubes. In spite of the recent advancement in X-ray tube technology, they still are limited in energy output to about 450 KeV. This energy level limits the penetration power to less than 10 cm of steel. For this reason, an increasing number of ports are nowadays turning to linac-based high-energy X-rays cargo scanners. These are based on the same concepts as the linac-based nondestructive testing applications discussed in the previous section of this chapter (Section 6.4). In cargo inspection applications, linacs are used to accelerate electrons to energies of 3 to 9 MeV capable of producing high-energy X-rays that penetrate deep and give the port inspectors information about the nature of the cargo. The denser the object, the more the attenuation of the projected X-ray is and the darker the image produced for such an object. Thus, the X-ray scanner reveals the basic shapes of objects and differentiates between different objects inside a cargo container. The optimal energy needed depends on the X-ray detector used, the size of the cargo container to be inspected, and the overall scanning system design. However, many cargo screening system manufacturers have found that 3 to 6 MeV linacs produce the best tradeoff for overall performance and cost [32]. If the energy of the linac is below 2 MeV, the X-ray penetration is low for a densely loaded container. On the other hand, linacs accelerating electrons to energies significantly above 9 MeV would require extensive shielding and could produce neutrons, an undesired by-product.

Another consideration in designing linacs that are used in cargo scanners is the choice of the rate at which the X-rays are emitted, given that the objects being inspected are moving with respect to the linac-based X-ray source. Because of power considerations, linacs used in this application are pulsed linacs and not continuous-wave (CW) linacs. The pulses are about 4 microseconds long and have adjustable pulse repetition frequency (PRF) between 50 and 500 pulses per second. Such relatively high PRF allows for reasonable speed of scanning and, hence, acceptable flow of cargo without compromising the quality of the images.

6.5.1 Scanning Units

There are two approaches for inspecting railcars and cargo containers in high-volume operations using the scanning unit based on linacs:

Stationary scanning units. In this inspection approach, the truck that is carrying the cargo container pulls up to a scanning station (which looks like a car wash construction) that has a linac, with the target producing the X-ray on one side and a detector on the other side. While the truck is moved slowly through the unit the fan of X-rays that has passed through the truck and its cargo is detected by a linear array of scintillation detectors. A transmission image is built up in real time, revealing what is inside the container. As an example, we show in Figure 6.10 a system built by Rapiscan Laboratories (Sunnyvale, CA). It is based on 6 MV X-ray linac imaging system that is capable of penetration up to 340 mm (13.9 in) in steel. This makes it possible to check the content of shipping containers or trucks with steel walls in the loading area. A linac producing higher energy X-ray would be able to reveal more clearly the contents of containers with more dense objects. An example is the Eagle F9000 system by Rapiscan, which is based on a 9 MV X-ray linac. Its imaging system is capable of penetrating more than 400 mm (16 in) of steel. In Figure 6.11 is shown a truck inside the stationary unit. The contents of the truck are revealed in Figure 6.12. Another example of the detected images is shown in Figure 6.13, which depicts a picture of a container concealing three smuggled cars detected by a fixed system [33].

Figure 6.10 A stationary inspection station based on the Rapiscan Eagle® F6000 System.

Figure 6.11 A truck being inspected inside a stationary unit (Rapiscan Eagle® F6000 System).

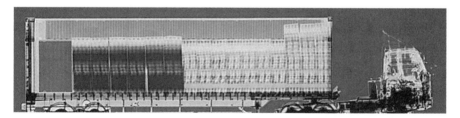

Figure 6.12 Contents of truck being inspected inside the stationary unit.

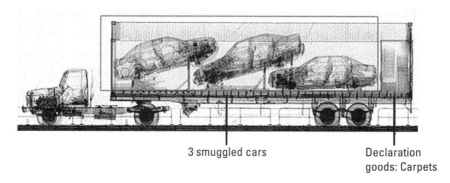

Figure 6.13 Three smuggled cars detected by high energy X-ray [33].

The use of a fan X-ray beam has several advantages. First, the X-rays that are scattered out of the fan by the cargo would miss the narrow detector, thus avoiding interference with the detected image. Second, spreading the beam into a fan makes the radiation dose low, protecting any stowaway from a dose more than the generally accepted radiation dose. Third, this radiation configuration makes it easier to design the shielding of the unit, resulting in low leakage levels outside the inspection tunnel and thus protecting the general public.

Mobile Scanning Units. In the second approach, a mobile scanning system is mounted on a special vehicle. During operation the mobile scanning system pulls right up to the truck that is carrying the cargo container being inspected. The mobile scanner unit moves up and down parallel to the container. The mobile scanner has the linac and its X-ray target on one side and a detector on the other side of the truck. An image is collected based on the X-rays that have passed through the truck and its cargo. Figure 6.14 shows a still truck being scanned by a moving scanning bridge.

6.5.2 Recent Advancements in Cargo Inspection

The cargo inspsction and ports security fields have benefitted from the advancement in imaging and pattern recognition techniuqes that have been developed recently. Most of these advanced techniques were originally developed for improved medical diagnostics and enhanced images for nondestructive testing. Examples of these techniques are the X-ray computed tomography, the use of dual energy X-ray systems, and combining these two techniques into dual-energy CT imaging. Additionally, recent advancements in the X-ray detectors have resulted in more sensitive, lower cost, and higher resolution imaging systems.

6.5.2.1 X-Ray Computed Tomography

X-ray computed tomography is widely used for medical diagnostic, NDT, and, more recently, for security purposes like baggage inspection. In this technique

Figure 6.14 A still truck being scanned by a mobile scanning unit.

cross-sectional images, or *slices*, of an object are numerically reconstructed from X-ray projections at various angles around the object. Using specialized computer programs, these cross-sectional images are combined to produce a two- or three-dimensional image (as in medical CAT scans). CT scanners measure the attenuation coefficient of the scanned object. The attenuation coefficient depends on the material being scanned and is also a function of the energy of the incident X-ray photons. In most current cargo inspection CT systems the scan is performed with an X-ray source of a specific energy spectrum, and it is not possible to determine the chemical composition of the scanned materials. Fortunately, some cargo inspection systems started recently to use dual energy X-ray systems. These systems yield material discrimination through the comparison of the attenuation of X-ray beams having two different energy spectra.

6.5.2.2 Dual-Energy X-Ray Systems

Recently, dual high-energy X-ray systems have been proposed for the detection of special nuclear material in cargo containers [33–35]. They use a linac that can be switched between two modes of operation to produce two different energies. In one mode of operation, 6 MeV electrons are generated producing, through the bremsstrahlung effect, X-rays with energies from 0 to 6 MV. In the other energy mode, the linac generates 9 MeV electrons, and thus producing X-rays with energy spectrum from 0 to 9 MV. Photons of two different energy levels interact with matter differently. The high atomic number (Z) materials are much more opaque to 9 MV photons than to 6 MV photons, so high Z materials, such as uranium and plutonium, can be recognized. Then, a special computer algorithm calculates the size and atomic numbers of individual objects and, based on that data, determines the level of concern and the possibility of the inclusion of such dangerous materials. We can get higher electron energy with higher RF power and lower electron current, and vice versa for lower energy X-rays. Thus, we can achieve the interlacing dual energy X-ray delivery by controlling and changing two variables, the injected beam current and the power of the RF source's (magnetron or klystron) from pulse to pulse. Figure 6.15 shows an RF scheme as an example for varying the electron beam energies from pulse to pulse [35]. However, an important requirement to achieve material identification is to have very stable dose rate X-ray pulses for both energies.

6.5.2.3 Dual-Energy CT Imaging

Dual-energy CT combines the above two advanced approaches into one technology, where the object is scanned with two different X-ray energy spectra. It allows not only the reconstruction of two-dimensional images of the objects in the container, but can also be used to estimate the density and the effective atomic number of different objects. In the security domain, these numbers,

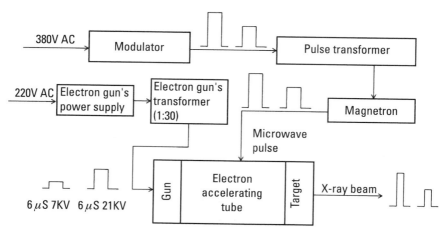

Figure 6.15 Scheme for interlaced dual energy X-ray system [35].

along with other features like volume and texture, can lead to accurate detection of explosives and dangerous materials, such as uranium and plutonium.

6.5.2.4 Advancement in the X-Ray Detector

The X-ray detector is where the photons, transmitted through the cargo, are converted into visible light (scintillation) and measured by sensitive photo-detector elements. The field of X-ray detectors for cargo scanning has benefitted significantly from advancements achieved by design and manufacturing engineers building detectors for medical diagnostics. Some of the improvement for cargo detectors [32] can be summarized as follows:

1. *Modular approach in design and manufacture.* In this approach the detector is comprised of arrays of smaller modules. This concept allows the designer to optimize each module for performance and at the same time achieve cost-effective manufacturability. Another benefit is being able to make the active area of the detector as small as possible, thus improving the detector's resolution.
2. *Improvements in sensitivity.* The sensitivity of the detector is a measure of the smallest change in signal that can be detected. The enhancement in detector sensitivity came as a result of the detector circuitry improvements that reduce the influence of noise and scatter on the quality of the detected signal.
3. *Larger dynamic range.* This parameter describes the ratio of the highest to the lowest signal that the detector can produce. Improvements in this detector parameter have resulted in more precise imaging for both empty and fully loaded containers.

6.6 Ion Implantation in Semiconductor Chip Fabrication

The use of accelerators in the manufacturing of integrated circuits represents a significant market for industrial accelerators that is comparable in size to the market of medical linacs [24]. The ion implanter is used to alter the near-surface properties of semiconductor materials [36]. Doping or otherwise modifying the silicon and other semiconductor wafers was done earlier by thermal diffusion of deposited dopants. During the 1960s improvements were made in the electrostatic accelerators such as the Van de Graff and some of its modified versions such as the Pelletron developed by Herb and collaborators, which became available from the National Electrostatics Corporation (NEC) [37]. These improvements and the progress in building compact electrostatic accelerators pioneered the development of commercial ion implanters. Gradually, the use of ion implanters for doping started to replace doping by thermal diffusion. Meanwhile, experience gained in building research accelerators improved the hardware reliability of ion implanters and generated new techniques for purifying and transporting their ion beams. Thus, by the early 1970s the ion implantation using electrostatic accelerators started to become the technique of choice for doping semiconductors. Among semiconductor processing techniques, ion implantation is nearly unique in that process parameters, such as concentration and depth of the desired dopant, are specified directly in the equipment settings for the dose and energy, respectively, of the implant ion.

Doping and modifying silicon and other semiconductor wafers by ion implantation encompasses generating an ion beam and steering it into the substrate so that the ions come to rest at a specific precise depth beneath the surface. Some of the most commonly implanted ions are arsenic, phosphorus, boron, indium, and antimony. The ion energies of ion implanters must be variable over a wide range to allow for different depths for different dopant ions. In Table 6.3, we list the common ranges of current and energy sorted in three categories: high-current, medium-current, and high-energy.

6.6.1 Concept of Operation

The ion implanter generates the appropriate dopant ions and selects, accelerates, and scans them. A simple scheme of an ion implanter is shown in Figure 6.16.

Table 6.3
Current and Energy Ranges of Ion Implanters

High-Current Ion Implanters		Medium-Current Ion Implanters		High-Energy Ion Implanters	
Ion Current	Energy	Ion Current	Energy	Ion Current	Energy
Up to 30 mA	1 keV to 200 keV	1 micro A to 5 mA	2 keV to 900 keV	up to 1 mA	up to 5 MeV

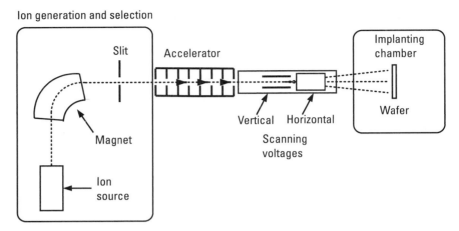

Figure 6.16 Schematic of the main constituents of a typical ion implanter.

In this figure we show the main constituents of the ion implantation system schematically, where the sizes of different subsystems are not in proportion, just for illustrative reasons.

In the ion, source gaseous species are ionized to produce the dopant ions. An analyzing magnet selects the specific ion to be implanted. The strength of the magnetic field of the analyzing electromagnet is set to provide the specific field that corresponds to the ion mass and charge. A slit allows only the chosen ions to pass to an electrostatic accelerator. At this step, the ions are accelerated to attain the desired kinetic energy commensurate with the nominal depth of implantation. In the scanner, the beam is scanned horizontally and vertically, and precisely in order, to provide a uniform dose of ions to the specific area to be doped on a silicon wafer.

6.7 Concluding Remarks

We reviewed briefly in this chapter some of the industrial applications of accelerator-based electron and X-rays beams, from the treated rubber for tires on our cars, to the sterilized medical disposables we use, to the computer chips that run our computers, to the securing of our ports all over the world. We make use of these applications in our daily lives and in many cases we may not be aware of the technology behind it. The use of accelerators in these applications provides the unique trait of allowing for treatment of objects in large volumes and at the same time providing precise control of the doses delivered.

In the next chapter we discuss facilities and applications using large accelerators. One group of these facilities is based on the production of synchrotron radiation. Beside the multiple scientific applications of this technology, it has proved to hold unique qualities that have been recently exploited in some important industrial applications. Examples of such fields are the pharmaceutical industry and the semiconductor fabrication Industry. Some of the applications of synchrotrons, including the current and potential ones, are discussed in the next chapter. In Chapter 8 we look at accelerator-based technologies that are at the frontier and hold the promise to shape the future of scientific as well as commercial accelerators. Two examples that have current and potential industrial accelerator-based applications are the free-electron lasers and neutron spallation sources.

References

[1] Berejka, J., and M. R. Cleland (eds.), "Industrial Electron Beam Processing," Revision 4, Vienna: International Atomic Energy Agency (IAEA), 2010.

[2] Cleland, M. R., et al, *Equations Relating Theoretical Electron Range Values to Incident Electron Energies for Water and Polystyrene*, RDI-IBA TIS 1556, Edgewood, NY: IBA Industrial Inc., 2003.

[3] Mevex Corporation, Stittsville, Ontario, Canada, http://www.mevex.com/index.html.

[4] Cleland, M. R., "Industrial Applications of Electron Accelerators," CERN Accelerator School, Zeegse, The Netherlands, May 24–June 2, 2005, http://cas.web.cern.ch/cas/Holland/PDF-lectures/Cleland/School-2.pdf.

[5] Hunt, J. D., and G. R. Alliger, "Application of Radiation to Tire Manufacture," *Radiation Physics and Chemistry*, Vol. 14, No. 1–2, 1979, pp. 39–53.

[6] Ashbaugh, C. E., III, "Gemstone Irradiation and Radioactivity," *Gems and Gemology*, Winter 1988, pp. 196–213, http://lgdl.gia.edu/pdfs/ashbaughw88.pdf.

[7] Berejka, A. J., "X-Ray Curing of Adhesives," *RadTech Report*, Nov.-Dec. 2006, pp. 18–21.

[8] L-3 Communications, http://www.l-3com.com/

[9] Saravacos, G. D., and A. E. Kostaropoulos, *Handbook of Food Processing Equipment*, Food Engineering Series, New York: Springer, 2002.

[10] Hall, E. J., and J. G Amato, *Radiobiology for the Radiologist*, 6th ed., Philadelphia: Lippincott William and Wilkins, 2006.

[11] Moran, J. P., "Cost-Effective Red Water Disposal by Electron Beam Radiolysis," Somerville, MA: Science Research Lab, Tech. Report SRL-10-F, 1994.

[12] Pikaev, T., "Environmental Applications of Radiation Technology," *High Energy Chemistry*, Vol. 28, No.1, 1994, pp 1–10.

[13] "Radiation Processing: Environmental Applications," Vienna: International Atomic Energy Agency, 2007.

[14] Cooper, W. J., Curry, R. D., and O'Shea, K. E. (eds.), *Environmental Applications of Ionizing Radiation*, John Wiley, 1998.

[15] Han, B., et al., "Electron Beam Treatment of Textile Dyeing Wastewater: Operation of Pilot Plant and Industrial Plant Construction," *Proc. of IAEA Tech. Meeting*, Sofia, Bulgaria, September 7–10, 2004, IAEA-TECDOC-1473, pp 101–110, http://www-pub.iaea.org/mtcd/publications/pdf/te_1473_web.pdf.

[16] High Voltage Environmental Applications, Inc., "High Voltage Electron Beam Technology–Innovative Technology Evaluation Report," EPA/540/R-96/504, August 1997.

[17] Sampa, M. H. O., et al., "Treatment of Industrial Effluents Using Electron Beam Accelerator and Adsorption with Activated Carbon: A Comparative Study," *Radiation Physics and Chemistry*, Vol. 71, 2004, 457–460.

[18] *Radiation Treatment of Gaseous and Liquid Effluents for Contaminant Removal, Proc. of IAEA Tech. Meeting*, Sofia, Bulgaria, September 7–10, 2004, IAEA-TECDOC-1473, http://www-pub.iaea.org/mtcd/publications/pdf/te_1473_web.pdf.

[19] Chmielewski, A. G., "Application of Ionizing Radiation in Environmental Protection," *Proc. of IAEA Tech. Meeting*, Sofia, Bulgaria, September 7–10, 2004, IAEA-TECDOC-1473, pp 11–24, http://www-pub.iaea.org/mtcd/publications/pdf/te_1473_web.pdf.

[20] Chmielewski, A. G., and J. Licki, "Electron Beam Flue Gas Treatment Process for Purification of Exhaust Gases with High SO_2 Concentrations," Nukleonika, Vol. 53, 2008, pp 61–66, http://www.nukleonika.pl/www/back/full/vol53_2008/v53s2p061f.pdf.

[21] "Radiation Processing of Flue Gases: Guidelines for Feasibility Studies," IAEA-TECDOC-1189, Vienna: IAEA, December 2000, http://www-pub.iaea.org/MTCD/publications/PDF/te_1189_prn.pdf

[22] Hellier, C. J., *Handbook of Nondestructive Evaluation*, New York: McGraw-Hill, 2001.

[23] Nunan, C. S., "Present and Future Applications of Industrial Accelerators," Ninth Fermilab Industrial Affiliates Roundtable on Applications of Accelerators, 1989, SLAC eCONF C8905261, pp. 55–82, http://www.slac.stanford.edu/econf/C8905261/pdf/004.pdf.

[24] Hamm, R. W., "Industrial Accelerators," Reviews of Accelerator Science and Technology, Vol.1, 2008, pp 163–184.

[25] Schonberg, R. G., et al., "Portable X-band Linear Accelerator Systems," *IEEE Trans. on Nuclear Science*, Vol. NS-32, No. 5, 1985, pp. 3234–3236.

[26] Tang, C., "The Development of Accelerator Applications in China," *Proc. of APAC*, Gyeongju, Korea, 2004, pp. 528–532.

[27] Yamamoto, T., et al., "Design of 9.4 GHz 950 keV X-Band Linac for Nondestructive Testing," *Proc. of EPAC 2006*, Edinburgh, Scotland, pp. 2358–2360.

[28] Hanna S. M., "Review of Energy Variation Approaches in Medical Accelerators," *Proc. of the 11th European Particle Accelerator Conf.*, Genoa, Italy, June 2008.

[29] E. Tanabe. Variable energy standing wave linear accelerator structure. US Patent 4286192, issued August 1981.

[30] R. H Giebeler. Accelerator side cavity coupling adjustment. US Patent 4400650, issued August 1983.

[31] S. M. Hanna. Electronic energy switch for particle accelerators. US Patent 7112924 B2, issued September 2006.

[32] Reed, W. A., "X-Ray Cargo Screening Systems: The Technology Behind Image Quality," *Port Technology International*, PT35–13-1, 2007, pp. 1–2.

[33] Wang, X., et al., "Material Discrimination by High-Energy X-ray Dual-Energy Imaging," *High Energy Physics and Nuclear Physics,* Vol. 31, No. 11, November 2007, pp. 1076–1081.

[34] Chen, Z., and X. Wang, "Cargo X-ray Imaging Technology for Material Discrimination," *Port Technology International Customs and Security,* PT 30-41-2, 2007, pp. 1–3.

[35] Tang, C., H. B. Chen, and Y. H. Liu, "Electron Linacs for Cargo Inspection and Other Industrial Applications," *Proceedings of International Topical Meeting on Nuclear Research Applications and Utilitzation of Accelerators,* SM/EB-28, IAEA, Vienna, Austria, May 2009, pp. 1–8.

[36] May, G. S., and C. J. Spanos, *Fundamentals of Semiconductor Manufacturing and Process Control,* Hoboken, NJ: John Wiley and Sons Inc., 2006, pp. 56–58.

[37] Herb, R. G., "Early Electrostatic Accelerators and Some Later Developments," *IEEE Trans.,* NS-30, No. 2, 1983, pp. 1359–1362.

7

Large Accelerators

Almost all the advancements in the accelerator technology originated from the research and development achieved in national laboratories and research centers developing the larger and more complex scientific accelerators. In this chapter, we will discuss examples of large accelerators in three areas:

1. Gigantic beam colliders built by high-energy physicists to reveal the fundamental forces and particles of nature;
2. Synchrotrons as user-facilities for biologists, physicists, and chemists;
3. Particle therapy circular machines developed for precise cancer proton and ion radiation therapy.

7.1 Large Accelerator Facilities for High-Energy Physics

From the beginning and for almost a century now, the motivation for building large accelerators has been to accelerate particles to higher and higher energies in order to probe deeper and deeper into the structure of matter [1]. It starts by accelerating electrons or protons to smash them into a fixed *target*. The target is surrounded by a large detector that measures the emission angle, the energy, and momentum of the particles emitted from the target as a result of the collision. From simple physics, one would realize that we get more impact if we have two opposite particle beams colliding in a *head-on collision*. In this case all the initial particle energy is available for creating new particles and giving them momentum.

There are two basic configurations for *colliders*. In the *linear collider,* the two opposite beams are accelerated through sections of linear accelerators. In the *circular collider*, the two beams are accelerated in circular accelerators, such as a synchrotron. In this section we describe briefly the distinctive features of three examples of high-energy colliders that reveal recent developments in accelerator technology.

7.1.1 Linear Colliders

Linear colliders accelerates beams collinearly in opposite directions before colliding them in a specific spot called the *interaction point* (IP). This configuration mitigates some of the issues associated with circular colliders such as the energy loss due to synchrotron radiation. This loss is more notable for circular electron colliders where the orbiting electrons lose a lot of energy in synchrotron radiation (as will be discussed later this chapter). For this reason, linear colliders are better configurations for smaller particles, such as electrons and their antiparticles; positrons and circular colliders are preferably used to accelerate heavier particles, such as protons or ions (both of which are part of a group called *hadrons*), since the associated synchrotron radiation loss is less than that for circulating electrons.

One example of linear colliders is the *Stanford Linear Collider* (SLC), which was the first linear collider of its generation [2, 3]. It was built in 1989 utilizing the already existing SLAC two-mile linear accelerator [4]. The SLC operated for nine years until 1998 and produced important high-energy physics results. The SLC is a folded version of a linear collider, in which bunches of both electrons and positrons (antielectrons) are accelerated in a single pass through the linear accelerator to 50 GeV each. At the end of the two-mile linac, they are bent into two separate semicircle arcs (see Figures 7.1, 7.2). The arcs are 600-meter loops that bend the beams of electrons and positrons using sets of magnets. The oppositely charged particles are then brought to a head-on collision at the IP located in the center of a very large particle detector with a center of mass energy of 100 GeV.

In a collider such as the SLC, the probability for a particular process varies with what is known as the *luminosity*—a figure of merit quantifying the performance of a collider concerning its rate of collisions. It depends on the beam intensities, beam transverse cross-section, and repetition rate. For higher luminosity, we need to squeeze the maximum number of particles into the smallest amount of space around the interaction region. To increase the luminosity, one must either increase the beam intensity or decrease the beam size. SLC features very small beam spots, less than 2 microns across in the transverse directions at the IP; its repetition rate is 120 pulses per second, and the beam hosts 4×10^{10} particles per bunch [5]. This allows enough luminosity for the SLC's experiments to contribute effectively to most of the high-energy physics research interests it was built to explore.

Figure 7.1 Aerial view of the Stanford Linear Collider.

7.1.2 Circular Colliders

In a circular collider, two beams of particles travel in opposite directions in a ring-shaped accelerator (called the *storage ring*). The circulating beams are brought to collision at an IP in the middle of a large detector that senses the results of the collision.

One important limitation of circular colliders is the associated loss of energy. A circulating charged particle beam emits energy in the form of electromagnetic radiation called *synchrotron radiation*. As the velocity of the circulating charged particle approaches that of light, the loss of energy by the particle multiplied by the number of particles in the beam rises sharply, making the beam lose enormous power that can reach megawatts. The amount of synchrotron radiation power P emitted in a ring of fixed radius r scales as the fourth power of the beam energy U and inversely as the third power of the beam particle's mass m, as shown in (7.1).

$$P \sim \frac{U^4}{r^2 m^3} \tag{7.1}$$

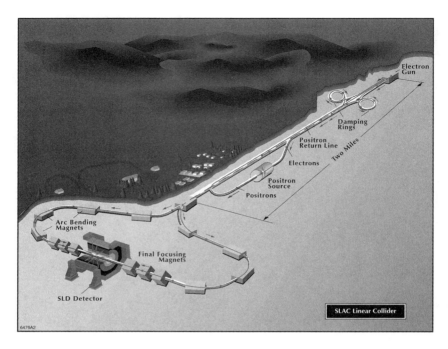

Figure 7.2 Simplified schematic for the Stanford Linear Collider.

Thus, for electrons, which are about 1,800 times lighter than protons, synchrotron radiation becomes a severe constraint. To limit energy loss due to synchrotron radiation, the bending radius of an electron storage ring must increase approximately linearly with energy, and thus the size and cost increase as the energy squared. This constraint dictates that high-energy electron circular colliders must have a very large and gentle radius. An example is the circular electron-positron (e^- − e^+) collider built by the *European Organization for Nuclear Research* (CERN) and discussed briefly below.

7.1.2.1 Large Electron Positron Collider

The *Large Electron Positron* (LEP) synchrotron collider was built in 1989 at CERN, a multinational center for research in nuclear and particle physics near Geneva, Switzerland. LEP was a circular collider with a circumference of 27 kilometers (17 miles) built in a tunnel 100 meters deep and straddling the border of Switzerland and France [6, 7]. The reason for such a large radius is to reduce the synchrotron radiation loss that, as shown in (7.1) above, is inversely proportional to the square of the orbit of the circulating particles. The power and cost of the RF system is determined to a large extent by the need to replace the energy lost by the particles due to synchrotron radiation. The first beam collisions at LEP took place in summer 1989 at beam energy of 46 GeV.

The original RF accelerating system used room-temperature copper cavities. To reduce the electrical power consumption and to be able to upgrade to higher colliding energies, LEP added superconducting RF accelerating cavities. In the upgraded LEP machine, electron and positron beams, each with energy of 100 GeV, traveled around the circular tunnel to achieve a collision with a center of mass energy of 200 GeV.

LEP was considered the most powerful electron accelerator ever built until now. It was used for eleven years, from 1989 until 2000, and it was later decommissioned to make room in the tunnel for the construction of the Large Hadron Collider (LHC).

7.1.2.2 Large Hadron Collider

The *Large Hadron Collider* (LHC) is a circular collider near Geneva, Switzerland [8]. Its name reflects its main three features: *Large* (its size is approximately 27 km in circumference); *Hadron* (because it accelerates protons and antiprotons); and *Collider* (because these particles form two beams travelling in opposite directions and collide at specific interaction points). The European Organization for Nuclear Research (CERN) built the HLC with the intention of testing various predictions of high-energy physics. It uses the tunnel used by LEP collider, discussed briefly above. The LHC spans the border between Switzerland and France, about 100 m underground (see Figure 7.3). It is the world's largest and highest-energy particle accelerator built so far. This synchrotron is designed to collide opposing particle beams of protons and antiprotons at energy of 7 teraelectron volts (7 TeV; one TeV is 10^{12} electron-volts). It is expected to address some of the most fundamental questions of physics, advancing the understanding of the deepest laws of nature. Its size is related to the maximum energy obtainable. As with all circular colliders or storage rings, the energy is a function of the radius of the machine and the strength of the magnetic field that keeps particles in their orbits.

After operating for a few days in September 2008, a serous fault caused this gigantic machine to shut down. It resumed operation in November 2009 and demonstrated its first planned collisions between two 3.5 TeV beams, a new world record for the highest-energy man-made particle collisions. The LHC will continue to operate at half power for some years; it will not be running at full power (7 TeV per beam) until 2014.

As in any collider, the most important parameters are the beam energy and its luminosity. Similar to our discussion of linear colliders, luminosity of a circular collider such as the LHC is used to indicate the probability for a particular process. It depends on the number of particles in each bunch, the frequency of complete turns around the ring, the number of bunches, and the beam cross-section. Basically, we need to squeeze the maximum number of particles into the smallest amount of space around the interaction region. In the

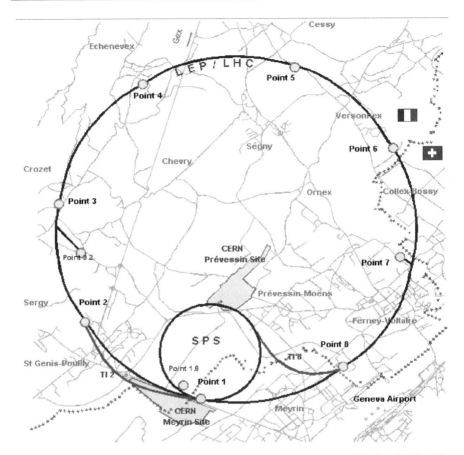

Figure 7.3 Map of the LHC crossing the borders between France and Switzerland, with a circumference of 27 km.

LHC, under nominal operating conditions, each colliding beam will consist of nearly 3,000 bunches of particles, and each bunch will contain as many as 100 billion particles! The particles are so tiny that the chance of any two colliding is very small. When the bunches cross, it is predicted that there will be only about 20 collisions between 200 billion particles. Bunches will cross on average about 30 million times per second. Thus, the LHC will be able to generate up to 600 million particle collisions per second. Some of the major design parameters of the LHC are listed in Table 7.1, below.

7.1.3 Use of Superconductivity in Large Accelerators

The linac applications reviewed in Chapters 5 and 6 are all based upon accelerators using copper cavities. In large accelerators, where large RF power is needed

Table 7.1
Basic Parameters of the LHC

LHC Parameters	
Circumference	26.659 km
Number of magnets	9,593
Number of RF cavities	16
Nominal energy, protons	7 TeV
Peak magnetic dipole field	8.33 T
Min. distance between bunches	~7 m
Design luminosity	10^{34} cm^{-2} s^{-1}
No. of bunches per proton beam	2,808
No. of protons per bunch (at start)	1.1×10^{11}
Number of turns per second	11,245
Number of collisions per second	600 million

to accelerate the charged particles to high energy, a large amount of power is dissipated in the ohmic resistance of the copper cavity walls. The cost of generating this additional power and also disposing this power through cooling is a dominant part in the power budget of the operation of any large accelerator and, hence, a major cost for any of these large accelerators.

Since the resistance of a normal conductor increases as its temperature rises, one would think that the solution is to refrigerate the copper cavities to cryogenic (extremely low) temperatures. However, the resistivity of cryogenically cooled copper is reduced by less than an order of magnitude. From a power budget point of view, this is not practical, since the power used to refrigerate the copper cavities is more than the power saved by reducing the copper resistance by cooling.

A more practical approach is to build high-power RF cavities of a superconductor material such as niobium [9, 10]. Superconducting cavities are characterized by having small energy losses and large stored energy. In a superconductor below a certain temperature called the *transition temperature* (Tc), there is virtually no resistance to the current flowing in such a material. This remarkable property of superconductivity is attributed to the fact that in a superconductor below Tc, and below a *critical magnetic field*, electrons bind in pairs known as *Cooper pairs*. At low temperatures these electron pairs contribute to the moving of electric charge freely within the lattice of atoms in the superconducting material. Thus, the resistance of a superconducting material can drop six orders of magnitude between the Tc of the material and zero temperature, as shown in Figure 7.4 [11].

Superconducting cavities have found successful application in a variety of large accelerators spanning a wide range of accelerator requirements. Superconductor cavities are commonly used in high-luminosity, high-energy physics

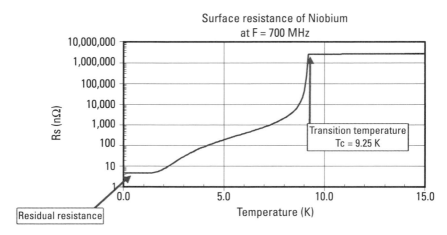

Figure 7.4 Temperature variation of the surface resistance of the superconducting material niobium at a frequency of 700 MHz.

colliders, as well as high-current storage rings for synchrotron light sources. CERN moved to superconducting RF cavity when it upgraded the LEP collider to its second phase. Three hundred superconducting four-cell RF cavities were built for this application. To build 300 such units there were considerable savings in material cost by fabricating the cavity out of copper and coating it with niobium by sputtering. The frequency of choice was 350 MHz. Figure 7.5 shows a typical 4-cell module [12].

Figure 7.5 A 4-cell, 350 MHz Nb-Cu cavity used in the upgrade of LEP.

7.2 Synchrotrons Sources

In previous chapters, we covered many applications for X-rays. In Chapter 5 we discussed the use of X-rays in cancer radiation therapy (RT). We have reviewed some of the very diverse industrial applications for X-rays in Chapter 6. In this section, we look at X-rays as one of the main probes that extend our current capability to analyze molecular properties of materials. This is attributed to the availability of *synchrotron radiation* with its exceptional properties.

7.2.1 Synchrotron Storage Rings

Synchrotron radiation was seen for the first time by the researchers at General Electric in the United States in 1947 in an accelerator they called the *synchrotron*, and hence the name coined for this type of radiation. It was first considered a nuisance because it caused the particles to lose energy in synchrotrons used for colliders. However, soon after it was recognized as a useful electromagnetic radiation, usually called the *synchrotron light*, with exceptional properties. Tens of synchrotrons have been built around the world solely as sources for this type of radiation, extending from infrared to hard (short wavelength) X-rays. The generic name *light source* is used often to describe these user facilitates, where scientists from universities and industry use this powerful tool for research in many fields.

The synchrotron is a circular accelerator (also called a *storage ring*) that guides charged particles, such as electrons, into an orbit at almost the speed of light. When electrons are deflected through magnetic fields, they emit high-intensity electromagnetic waves that are produced as a high-flux narrow beam. An electron circulating in the ring is thus like a moving torchlight, producing a narrow photon beam in the direction tangential to the ring (see Figure 7.6). Typically, several *beamlines* are installed around the storage ring to receive the

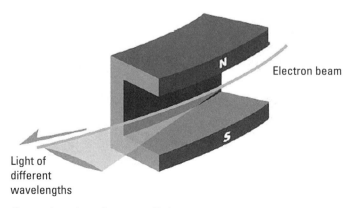

Figure 7.6 Generation of synchrotron radiation.

resulting radiation (see Figure 7.7; note that the electron injector is not shown, for simplicity).

In a typical user-facility synchrotron, different experimental setups are normally established along each beamline to carry out material investigations exploiting the unique properties of synchrotron radiation. In Figure 7.8, we show an example of layout of a synchrotron (Lawrence Berkeley National Laboratory). In Figure 7.9, we show a beamline in the same synchrotron, with the typical associated experimental setups [13].

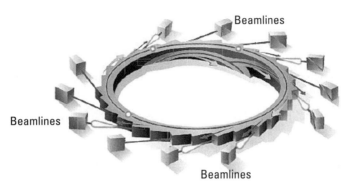

Figure 7.7 Beamlines around a synchrotron.

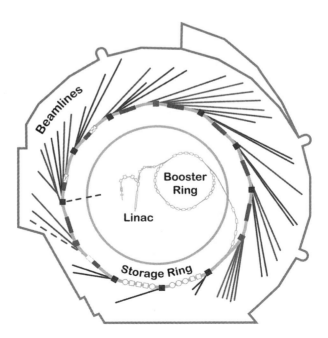

Figure 7.8 A layout of a synchrotron (Lawrence Berkeley National Laboratory).

Figure 7.9 A typical synchrotron beamline with scientist establishing experimental setups to use the synchrotron radiation.

Let us now consider how this radiation is produced. When electrons are accelerated, whether they are gaining speed along a straight line or traveling at a constant speed on a curved path, they emit electromagnetic waves (see Figure 7.6). This process has the same basic principal behind the mechanism by which radio waves are emitted from an antenna. Electrons in an antenna of a radio emitter are oscillating, and their velocities are changing with time, thus radiating electromagnetic waves (radio waves in this case). Similarly, in a synchrotron, bunches of electrons being accelerated (changing their direction as they are constrained in the circular pass in a synchrotron ring) emit high-intensity photons. However, these two similar mechanisms differ in that electrons oscillating in a radio antenna emit basically one frequency, or a single wavelength. However, in the case of synchrotron, electrons circulating in the ring illuminate each beamline only for a very short period of time, producing a *short pulse* in the form of a burst of photons. A short pulse in time corresponds to a wide band of frequencies. In the case of synchrotron radiation, a wide band of frequencies also means a wide band of wavelengths from which any desired wavelength can be selected with special filtering devices call *monochromators*. Figure 7.10 shows how a beamline is used to transport photons to the sample under test and the spectra filtering using a monochromator [13, 14].

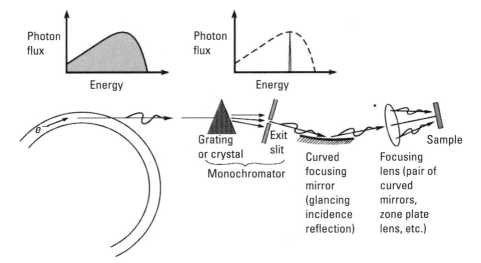

Figure 7.10 Transporting and filtering synchrotron radiation from a storage ring.

Synchrotron radiation produces photons with wavelengths that extend from the infrared to hard X-rays with energies of hundreds of keV with uniquely high photon intensities. The part of the electromagnetic spectrum that synchrotron radiation can occupy is shown in Figure 7.11. Note that VUV stands for *Vacuum UV*, and it describes a range of UV with wavelengths around 100 nanometer and of energy in the neighborhood of 10 eV. It is so named because it is absorbed strongly by air and is, therefore, used only in vacuum.

7.2.2 Wigglers and Undulators

Early synchrotron radiation facilities were basically circular rings where the radiation resulted from the bending of the electron beams under the effect of the bending magnets. They generally had an electron beam of relatively large cross-section and angular divergence. Modern user-facilities are dedicated to

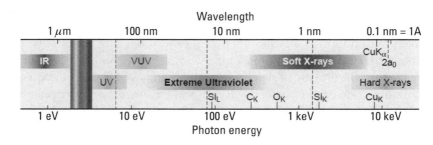

Figure 7.11 Portion of the electromagnetic spectrum that synchrotron radiation occupies.

broad scientific uses that are optimized to offer collimated beams with high brightness. This is done by the inclusion of many long, straight sections fitted with *wigglers* and *undulators*.

Undulators are periodic magnetic structures producing sequences of alternating, relatively weak, magnetic fields—that is, north-south, south-north, and so on. This periodicity causes the electron to experience oscillations as it moves along the axis of the device. The resulting motion is characterized by small angular excursions called *undulations* [13]. Since the magnetic field is relatively weak, the amplitude of the undulations is small, resulting in a narrow radiation cone as shown in Figure 7.12. An undulator produces X-ray with a few orders of magnitude in brightness more than the synchrotron light which are produced in conventional synchrotron storage rings by bending the electron beam using bending magnets, as discussed above in Section 7.2.1.

Wigglers are also periodic magnetic structures, the same as undulators, but they are characterized by a relatively stronger magnetic field. They produce radiation cones with broader angles than those produced by undulators. Another difference between the undulator and the wiggler is that the undulator results in a longer pulse of light, rather than a series of short bursts, as in the wiggler. The longer pulses are characterized by the emission of photon energies concentrated over short bands of wavelengths. In constrast, the wiggler produces photons that have a continuum of wavelengths similar to that of bending magnet radiation, but increased by the number of magnetic pole pieces. A photo of a 2-meter-long wiggler with 55 poles of permanent magnets from Stanford is shown in Figure 7.13 [15].

7.2.3 Examples of Synchrotron Radiation Applications

As was mentioned above, the high-intensity synchrotron radiation produces photons with wavelengths that extend from the infrared to hard X-rays, with

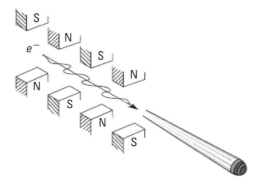

Figure 7.12 Radiation beam resulting from undulating effect.

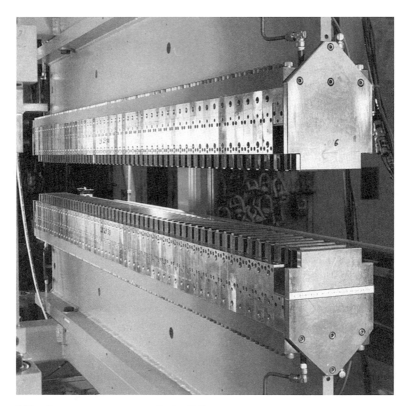

Figure 7.13 A permanent magnet wiggler [15].

uniquely high-photon intensities that can be used to determine the composition and properties of materials using a variety of techniques. The fields of material and chemical science have benefited greatly from the application of synchrotron radiation–based analytical methods [16, 17]. These include research in the areas of materials chemistry, surface science, magnetic materials, and chemical dynamics. Samples of useful applications include the magnetic properties of hard discs for memory storage, wear of aircraft turbine, and the study of catalysts in chemical processes. Synchrotron radiation can be used to measure atomic separation to a hundredth of an angstrom. In any material, adjacent atoms are bonded by the action of outermost electrons orbiting around each atom. The energies available from synchrotron radiation are well suited to studying those bonds. For example, industries developing corrosion-resistant surfaces or catalytic converters would be interested in analyzing the bonding mechanisms of different gases with metallic surfaces. Another example is the use of *X-ray diffraction* technique in studying the crystallographic structure and physical properties of materials and thin films. This technique is based on observing the scattered intensity of an X-ray synchrotron beam hitting the sample as a function of incident and scattered angles, polarization (orientation of the electromagnetic fields in the wave), and wavelength. A third analytical technique is based on analyzing the X-ray emitted from a material when subjected to strong synchrotron radiation. It is called *synchrotron-radiation-induced x-ray emission* (SRIXE). This technique is used to determine elemental concentrations through the detection of characteristic X-rays produced by the interaction of the synchrotron X-ray beams with the sample material.

There are many ways that synchrotron radiation is used for study of biological materials. The most prominent example is the application of X-ray diffraction methods for determination of protein and DNA structures and understanding biological structure and function. This is an extremely active field of research pursued at many synchrotron facilities by scientists from universities, as well as industrial and government laboratories. Multiple synchrotron facilities have been involved in the Human Genome Project (a project to determinate the base sequence of the human genome).

Other examples include the use of synchrotron radiation in the design of drugs and in the configuration of enzymes for various industrial processes. One technique used is to crystallize complex biological molecules and then scatter the powerful synchrotron beams of X-rays off the atoms in the crystal to reveal its three-dimensional structure. This technique is known as macromolecular X-ray crystallography [18]. This technique has opened the door for gaining detailed knowledge of macromolecular structure and for probing large, complex biological molecules. The availability of such new analytical tools is providing a new approach for drug design by the pharmaceutical industry.

The manufacturers of integrated circuits face a challenging analytical problem since these integrated circuits are three-dimensional structures with features that have submicrometer dimensions. One of the effective techniques meeting this challenge is *computed microtomography* (CMT) using synchrotron radiation. It is a nondestructive technique that can be used to determine the microstructure of materials in integrated circuits with spatial resolutions of the order of 1 micrometer.

7.3 Cancer Particle Therapy

We have discussed in Chapter 5 conventional radiation therapy based on the use of linacs in treating cancer by producing high-energy electron or photon (X-ray) beams. In this section, we deal with cancer *particle therapy* (PT) [18–22]. The ionizing radiation in PT is accomplished by beams of particles heavier than electrons, such as protons or ions (mostly carbon ion), and employs relatively large circular accelerators. In general, the heavier the particle, the greater is its effectiveness in ionizing the tissue it traverses. For example, the proton has a mass 1,800 times the mass of an electron, and hence, a beam of protons would be more effective in destroying a cancerous tissue. Additionally, protons and carbon ions can target tumor tissue much more precisely than X-rays, possibly allowing the delivery of higher doses of radiation to a tumor. The majority of PT facilities are using protons. (Actually, the acronym *PT* is also sometimes used for *proton therapy*. An alternative name is *hadron therapy*, since *hadron* would cover both proton and ions.) Investigating different ions for cancer PT has shown that carbon ions represent a good alternative to protons. It has sharper stopping points and is more effective in killing malignant cells [21].

7.3.1 Advantages of Particle Therapy

In radiation therapy, proton beams differ from photon beams mainly in the way they deposit energy in living tissue. Whereas photons deposit energy in small packets all along their path through tissue, protons (or ions) undergo little scattering when penetrating in matter and deposit less energy along the way and give the highest dose near the end of their range (called the *Bragg peak*), just before coming to rest (Figure 7.14).

Thus, the depth-dose curves of proton and carbon ion beams are different from those of photons. After a short build-up region, conventional X-ray radiation shows an exponentially decreasing energy deposition, with increasing depth in tissue. In contrast, protons (and hadrons in general) show different depth-dose distribution characterized by delivering much of their energy to tissue just over the last few millimeters of their range, at the Bragg peak (Figure 7.15). It is clear from comparing the depth-dose curves of proton and carbon ion to those of typical conventional RT that protons and carbon ions can target

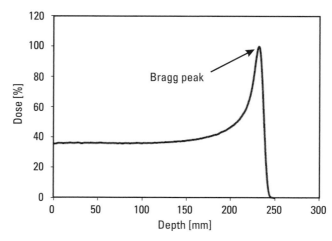

Figure 7.14 Typical dose deposition as a function of depth for a proton beam.

tumor tissue much more precisely than X-rays and penetrate deeper into the body with less harm to healthy tissue it traverses before and past the tumor site. This characteristic of protons results in an obvious advantage of proton treatment over conventional radiation because the region of maximum energy deposition can be positioned within the target for each beam direction. Changing the energy of the penetrating proton beam can change the precise position of the Bragg peak. The protons will stop at a particular depth, dependent on their inital energy.

Thus, it is possible to adjust this dose peak to within tenths of millimeters. Actually, the position of the Bragg peak is so sharp that it is generally smaller than the tumor. Consequently, it is possible to cover the tumor volume with high accuracy by varying the energy during the irradiation in a controlled way and superimposing many narrow Bragg peaks, to obtain a *spread-out Bragg peak* (SOBP). This is shown in Figure 7.16, with depth-dose distributions for an SOBP, its constituent pristine Bragg peaks, and a conventional RT X-ray beam (10-MV photon beam). The dashed lines indicate the clinical acceptable variation in the plateau dose of ±2%, showing the achievable uniformity of dose

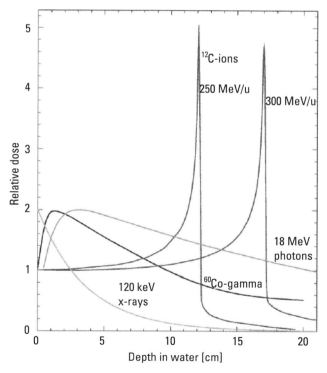

Figure 7.15 Depth-dose curves of proton and carbon ion as compared to conventional RT [22].

delivery over the area covered. Also an important feature is the absence of dose beyond the distal fall-off edge of the SOBP. Clearly, this is not the case with the photon beam.

Due to their relatively large mass, protons have little lateral side-scatter in the tissue; the beam does not broaden much, stays focused on the tumor shape, and delivers small dose side effects to surrounding tissue. This feature, and the fact that very few protons penetrate beyond the Bragg peak range, make proton therapy of particular interest for those tumors located close to critical organs and tissues, where a small local overdose can cause serious complications. Examples would be tumors close to the spinal cord, eye, or the brain, as well as in pediatric tumors.

7.3.2 Particle Therapy Circular Accelerators

In contrast to the conventional radiation therapy, which uses relatively compact linacs, PT systems are huge and much more expensive. In order to reach deep-seated tumors, proton or carbon ion beams must be accelerated to high enough energy. For example, to treat a tumor at a depth of 25 cm, the energy needed would be 200 MeV for a proton beam. Producing such energetic beams requires the use of large circular accelerators. PT facilities use either cyclotron or synchrotrons as accelerators. Each of these two approaches has its advantages. We will discuss both and point out a hospital-based facility as an example for each type of circular accelerators.

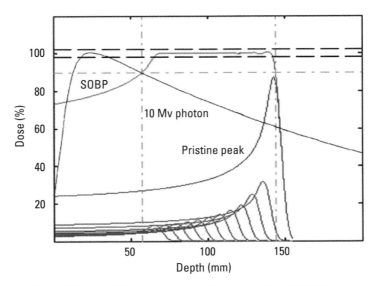

Figure 7.16 The SOBP dose delivery as compared to conventional RT [19].

7.3.2.1 Cyclotron Accelerators for PT

As explained briefly in Chapter 1, a cyclotron is a circular accelerator that accelerates charged particles in a spiral path. It consists of two semicircular electrodes (called the *dees*). In the center of the machine, an ion source emits a beam of ions or protons. The protons move inside the dees in a spiral trajectory under the combined effect of an electromagnet designed to produce a region of uniform dc magnetic field and the RF electric field produced across the gap between the dees by an oscillating voltage. As they spiral around, particles gain energy. Thus, they trace a larger arc, with the consequence that it always takes the same time to reach the gap. The size of the magnets and the strength of the magnetic fields limit the particle energy that can be reached by a cyclotron.

As an example of hospital-based proton treatment center using a cyclotron, we consider the Northeast Proton Therapy Center (NPTC) of the Massachusetts General Hospital (MGH) in Boston, Massachusetts. It is a cyclotron-based system manufactured by Ion Beam Applications (IBA). The cyclotron (*Cyclone 235*) delivers protons with energy of 235 MeV. A photo of the *Cyclone 235* is shown in Figure 7.17. The cyclotron produces the proton beam at a fixed energy. In order to provide energy variability to the system, the cyclotron is followed by an *Energy Selection System* (ESS). It is based on the use of a carbon stepped-wedge as an energy degrader, followed by a slit. The ESS transforms the fixed-energy beam extracted from the cyclotron into a variable-energy beam

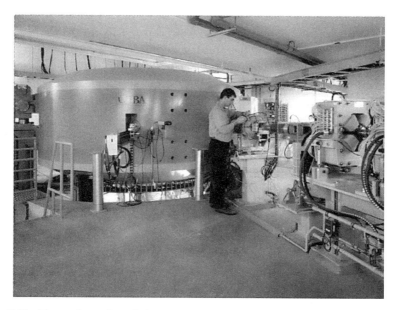

Figure 7.17 The cyclotron installed at the NPTC produced by IBA [24].

in the 235 to 70 MeV range. It can achieve a 10% energy variation within two seconds [24].

Although the cyclotron is only capable of providing accelerated protons for PT applications in fixed energy, it is generally a less expensive and a more compact option than the other PT alternative circular machine—namely, the synchrotron.

7.3.2.2 Synchrotron Accelerators for PT

The use of a synchrotron as the proton accelerator allows the production of proton beams with a variety of energies (unlike the cyclotron, which has fixed extraction energy). A small linear accelerator is often used to preaccelerate particles before they enter the ring. A synchrotron allows beam extraction for any energy, providing a more flexible solution.

The world's first hospital-based proton therapy center was commissioned in 1990 in the Loma Linda University Medical Center (LLUMC) [25–27]. The accelerator and the beam delivery system were designed by, and built with the help of, the U.S. Fermi National Accelerator Laboratory (Fermilab). A 250-MeV synchrotron was chosen as the proton accelerator. Protons are injected into the synchrotron at energy of 2 MeV. The synchrotron provides a pulsed beam to the designated treatment room every 2.2 sec. In the synchrotron storage ring, the protons gain energy as they pass through an RF cavity with each turn, until they reach the desired energy, ranging from 70 to 250 MeV, depending on the desired depth of penetration in the patient. At this point the magnetic dipole field and the RF frequency are held constant until the protons are extracted from the storage ring and transported to the treatment rooms. After extraction, the magnetic field returns to injection level for the next patch of protons.

The treatment facility at the LLUMC includes three gantries (one with robotic patient positioning) and one research room, as shown in Figure 7.18.

I discussed above two of the particle therapy facilities, one using a cyclotron and the other using a synchrotron. Although the operation of such facilities started only about two decades ago, this trend has been growing rapidly. As of May 2011, there are now 32 PT centers in operation and treating cancer patient around the world. Twenty-eight centers are treating with proton and four are using carbon ions. Up to that date, 73,804 cancer patients have been treated by proton beams and 6,661 patients treated by carbon ion beams [7, 28]. In spite of the size and cost associated with PT, it provides unique capabilities in treating specific case of cancer. As we discussed in this section, PT offers highly conformal treatment of deep-seated tumors with millimeter accuracy, while delivering minimal doses to the surrounding tissues. This distinctiveness makes this technique an effective cancer treatment that complements the conventional

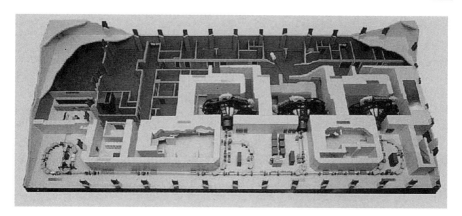

Figure 7.18 Facility layout of the proton center at the Luma Linda University Medical Center (LLUMC) [27].

radiation therapy based on linear accelerator to produce energetic electrons for both electron and X-ray therapies.

References

[1] Sessler, A., and E. Wilson, *Engines of Discovery: A Century of Particle Accelerators*, London: World Scientific Publishing Co., 2007.

[2] Emma, P., "The Stanford Linear Collider," *Proc. 16th IEEE Particle Accelerator Conf. (PAC-95)*, Dallas, TX, May 1–5, 1995, pp. 606–610.

[3] Seeman, J., "The Stanford Linear Collider," *Nuclear and Particle Science–Annual Review of Nuclear and Particle Science*, Vol. 41, 1991, pp. 389–428.

[4] Neal, R. B. (ed.), *The Stanford Two-Mile Accelerator*, New York: W. B. Benjamin Inc., 1968, pp. 245–247.

[5] Raimondi, P., et al., "Recent Luminosity Improvements at the SLC," SLAC-PUB-7847, July 1998, *6th European Particle Accelerator Conf. (EPAC-98)*, Stockholm, Sweden, June 22–26, 1998.

[6] Myer, S., "The LEP Collider: From Design to Approval and Commissioning," 91-08, Switzerland: CERN, 1991.

[7] Hübner, K., "Designing and Building LEP," *Physics Report 403–404*, 2004, pp. 177–188.

[8] CERN web site, "The Large Hadron Collider," http://public.web.cern.ch/public/en/LHC/LHC-en.html, last accessed January 2012.

[9] Padamsee, H., et al., *RF Superconductivity for Accelerators*, 2nd ed., Weinheim: Wiley-VCH Pub., 2007.

[10] Russenschuck, S., and G. Vandoni (eds.), "Superconductivity and Cryogenics for Accelerators and Detectors," CERN Report No. 2004–008, 2004.

[11] Sofa, H., "Surface Effects in SCRF Cavity," *Proceedings of the CERN Accelerator School*, Erice, Italy, May 2002, pp. 196–211.

[12] Padamsee, H., "Status of SRF Accelerator Technology," *SRF 2003 Workshop*, September 2003 http://srf2003.desy.de/talks/Padamsee/introSRF2003_v3.pdf, last accessed January 2012.

[13] Attwood, D. T., "Synchrotron Radiation for Materials Science Applications," University of California–Berkeley lectures, spring 2007, http://ast.coe.berkeley.edu/srms/2007/Intro2007.pdf.

[14] Attwood, D. T., *Soft X-Rays and Extreme Ultraviolet Radiation: Principles and Applications*, Cambridge University Press, 1999.

[15] SLAC National Accelerator Laboratory website, http://www.slac.stanford.edu/history/pix/ssr35.gif, last accessed January 2012.

[16] Jones, K. W., and H. Feng, "Microanalysis of Materials Using Synchrotron Radiation," Brookhaven National Laboratory, Report No. BNL-67588, 2000, http://www.bnl.gov/isd/documents/21473.pdf.

[17] Hirose, Y., "Materials Analysis using Synchrotron Radiation," *R&D Review of Toyota CRDL*, Vol. 38, No. 2, April 2003, http://www.tytlabs.co.jp/english/review/rev382epdf/e382_001hirose.pdf

[18] Helliwell, J. R., *Macromolecular Crystallography with Synchrotron Radiation*, Cambridge University Press, 1992.

[19] Lovin, W. P., et al., "Proton Beam Therapy," *British Journal of Cancer*, Vol. 93, 2005, pp. 849–854.

[20] Schlegel, W., et al., *New Technologies in Radiation Oncology*, Berlin: Springer-Verlag, 2006, pp. 345–363.

[21] Erickhoff, H., and U. Linz, "Medical Applications of Accelerators," *Review of Accelerator Science and Technology*, Vol.1, 2008, pp. 143–161.

[22] Amaldi, U., and G. Kraft, "Radiotherapy with Beams of Carbon Ions," *Rep. Prog. Phys.*, Vol. 68, 2005, pp.1861–18829.

[23] Weinrich, U., "Gantry Design for Proton and Carbon Hadron Therapy-Facilities," *Proc. of European Particle Accelerator Conf. (EPAC)*, June 27, 2006, Edinburgh, Scotland, pp. 964–968.

[24] Jongen, Y., et al., "Progress on the Construction of the Proton Therapy Center for MGH," *Proc. of the European Particle Accelerator Conf.*, Stockhom, Sweden, (EPAC), June 1998, pp. 2354–2356.

[25] Slater, J. M., et al., "Proton Radiation Therapy in the Hospital Environment: Conception, Development, and Operation of the Initial Hospital-Based Facility," *Review of Accelerator Science and Technology*, Vol. 2, 2009, pp. 35–62.

[26] Coutrakon, G. B., et al., "Performance Study of the Loma Linda Proton Medical Accelerator," *Med. Phys.*, Vol. 21, 1994, pp. 1691–1701.

[27] Patyal, B., "Maintenance and Logistics Experience at Loma Linda Proton Treatment Facility," http://ptcog.web.psi.ch/PTCOG49/presentationsEW/19-2-2B_Maintenance.pdf.

[28] Jermann, M., "Hadron Therapy Patient Statistics," Particle Therapy Cooperative Group (PTCOG), http://ptcog.web.psi.ch/Archive/Patientenzahlen-updateMay2011.pdf.

8

Recent Developments and Future Trends in Accelerator Technology

In this short chapter we look ahead at the current activities in the field of accelerators that are on the frontier and that are expected to be a major part of the future of scientific as well as commercial accelerators. Although these accelerators do not produce the highest energy beams and are not the largest in size, they are at the technology frontier and hold the promise to affect the future lives of many. In the previous chapter we covered briefly samples of "large accelerators." In this chapter we cover some of the accelerator technologies that are being developed or have been recently realized in a full scale. Some of these are still considered large accelerators, at least at this phase of development, such as the *free-electron laser* (FEL) Accelerators, covered in Section 8.1, and *Accelerator-Based Neutron Sources*, discussed in Section 8.2.

In all the linac applications that we have covered in this book so far, the basic force accelerating the particles is electromagnetic fields. In this chapter, we will touch on two different mechanism of acceleration. One comes from fields generated in plasma, called *plasma-based accelerators,* covered in Section 8.3. These *plasma wakefield acceleration* (PWFA) techniques could ultimately lead to powerful tabletop accelerators for science, medicine, and industry. The other novel acceleration mechanism is based on controlled discharge in transmission lines, as discussed in Section 8.4 on *dielectric wall accelerators*. One application being developed is building a proton therapy machine based on this technology.

The intent is to give the reader a quick glance at what is being developed—rather than a detailed and comprehensive coverage of these accelerator technologies—and to provide adequate references for readers who want to learn more about these new trends in the accelerator technology. Let us have a quick look.

8.1 Free-Electron Laser

Free electron lasers (FELs) represent an increasingly important kind of light source with a brightness that can be up to nine orders of magnitude (10^9 times) higher than that of ordinary synchrotron light (described in Section 7.2).

The basic principle of a free-electron laser is as follows [1]. Electrons are first brought to high energies in a linear accelerator. The tightly bunched beam of electrons is then fed into an undulator. As described in Section 7.2, the undulator is a long array of high-quality magnets with special arrangement of alternating polarity, as shown in Figure 8.1. When the electron beam passes through the undulator, magnetic fields of the undulator wiggle the electrons from side to side. This transverse acceleration causes the electrons to radiate photons in the forward direction similar to synchrotron radiation described in the last chapter. The higher the energy of electron beam, the higher the resulting photon energy. As electron bunches propagate down the undulator, they are bathed in the same light they generate. As they wiggle back and forth through the magnets and interact with the electric field of this light, some gain energy and some lose energy, depending upon their phase relationship with the light and the magnetic fields. As a result, the electrons begin to form *microbunches* separated by a distance equal to the wavelength of the light they generate. The light waves emitted by the electron bunches will line up in phase. This means that the waves' peaks and valleys overlay each other to reinforce and amplify the light's intensity coherently, thereby giving rise to light with properties characteristic of conventional lasers. By varying the energy of the linac electron beam, this laser wavelength can be varied over a wide range. The reason this photon source is called free electron laser is that the preceding process is similar to how the light from laser stimulates further emission of light within the laser medium (gas or solid). The difference is that this *lasting* effect in the case of the FEL is done in vacuum and there is no need for a physical medium for the electrons to amplify coherently the light produced—hence, the word *free* in the name FEL.

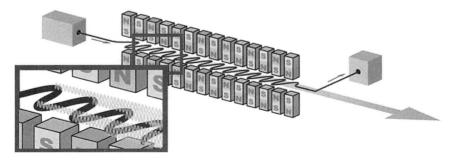

Figure 8.1 Simplified presentation of the operation of the FEL.

The FEL generates light in ultra-short pulses with the extremely high-peak intensity that can be used for strobelike investigations of extremely rapid processes. The FEL can be widely tunable in wavelength of the produced photons to cover frequencies ranging from microwaves, through terahertz radiation and infrared, to the visible spectrum, to ultraviolet, and to X-rays.

There are facilities that are built or currently being developed specifically as sources for high-brightness, high-coherence X-ray free electron laser (XFEL). Some of these facilities are in the United States, Europe, and Japan [2–4].

One example of the FEL new facilities that are designed specifically to produce FEL X-rays, is the *linac coherent light source* (LCLS) built at the Stanford University's SLAC National Accelerator Laboratory [2]. SLAC has converted its giant two–mile linac into the world's first X-ray laser, which started being available for research in late 2009. LCLS provides much shorter and much more intense X-ray pulses than ever available before. Scientists are now able to use the X-rays from FEL in observing chemical reactions, in which molecules join or split, that take place in a fraction of a nanosecond. The ultrafast LCLS X-ray flashes can capture images of such events with a "shutter speed" of less than 100 femtoseconds (1 femtosecond is one thousandth of a nanosecond or equals 10^{-15} of a second). LCLS is the first source to produce X-rays that are both very intense and clumped into ultrafast pulses. Some of the parameters of the SLAC's LCLS are listed in Table 8.1 [1].

From this brief review, we can see that the main aspects of the X-ray free-electron lasers are the extremely high-peak intensity combined with ultra-short pulse durations. One can put it simply that the FEL will allow the scientists to probing the ultra-small and capturing the ultra-fast.

Table 8.1
Some of the Parameters of the LCLS

Electron Beam	
Electron energy, GeV	14.3
Peak current, kA	3.4
Pulse duration, fs	67
Undulator	
Undulator period, cm	3
Undulator field, T	1.3
Undulator gap, mm	6.8
Total length, m	100
Magnet material	NdFeB
Radiation	
Radiation wavelength, nm	0.15
Bunches/sec	120
Peak power, GW	109

8.2 Accelerator-Based Neutron Sources

Every atom has in its nucleus neutral particles, the neutrons, and positively charged particles, the protons. Neutrons and protons have about the same mass, and both can be extracted from the nucleus and exist as free particles away from the nucleus. Neutrons are used as a research tool, such as in scattering experiments and other material analysis applications. Since neutrons can penetrate deeply into most materials, scientists use *neutron scattering* to probe the bulk of materials. Low-energy neutrons have energies of several tens of millielectronvolt (meV), corresponding to wavelengths of a few angstroms (note that particles can be viewed to act as waves under the right conditions). With these wavelengths, which are comparable to X-rays, neutrons scatter into diffraction patterns much as those seen in synchrotron radiation probing techniques. The two techniques often yield complementary information: neutron scattering (which is sensitive to the positions of the nuclei) and X-ray diffraction (which is sensitive to the positions of the orbiting electrons) are powerful probes of the structure of materials.

Neutrons have a greater sensitivity for light elements (elements with low atomic number), such as hydrogen and oxygen. One example of how scientists used this capability is their success in determining the critical positions of light oxygen atoms in yttrium-barium-copper oxide (YBCO), a promising high-temperature superconducting ceramic. Because neutron beams can be produced with energies that match the energies of atoms in motion, neutrons can be used to track molecular vibrations, movements of atoms during catalytic reactions, and changes in the behavior of materials subjected to outside forces, such as rising temperature, pressure, or magnetic field strength. Another useful characteristic of the neutron is that it has a magnetic moment (i.e., it acts as a very minute magnet). This property has been exploited in studying microscopic magnetic structure and developing magnetic materials and their applications, such as magnetic recording.

8.2.1 Neutron Spallation Sources

In general, *spallation* is a process in which fragments of material (spall) are ejected from a body due to impact. *Neutron spallation* is the process by which a particle accelerator is used to produce a beam of neutrons by impacting a target. Although it is more expensive way of producing neutron beams than by a chain reaction of nuclear fission in a nuclear reactor, it has the advantage that the beam can be pulsed with relative ease and in a well-controlled process.

Generally the production of neutrons at a spallation source begins with a high-powered linear proton accelerator, followed by a synchrotron to accumulate the proton beam bunches to produce an intense beam of protons that is then is focused onto a target [5, 6]. The target material should have high mass density to produce large neutron flux. It should also have high thermal conductivity to dispose easily the heat generated from the impact as well as high melting point (for solid targets). Solid targets of choice are tantalum or tungsten. Liquid mercury can be also used as target in neutron spallation. Spallation processes taking place in the target produce the neutrons, initially at very high energies. These neutrons are then slowed in what are called *moderators*, which are filled with liquid hydrogen or liquid methane. This step brings down the energies of the neutrons to the energies that are needed for the analytical techniques using neutrons.

There are already a few neutron source facilities that are operating around the world, such as the Spallation Neutron Source (SNS) [6, 7] in United States, ISIS [8] in United Kingdom, Swiss Spallation Neutron Source (SINQ) [9] in Switzerland, and there are a few in the pipeline, such as the Japan Proton Accelerator Research Complex (J-PARC) [10] in Japan and the European Spallation Source (ESS) [11] to be built in Sweden.

We will consider here the Spallation Neutron Source (SNS) [6, 7] as an example of neutron source facilities.

8.2.2 The Spallation Neutron Source (SNS)

The Spallation Neutron Source (SNS) is built on 80 acres in Oak Ridge, Tennessee (Figure 8.2). It is managed by Oak Ridge National Laboratory. It became the most powerful neutron source in the world in September 2009 by achieving a new world record when 1 MW of proton beam power was delivered to the target.

The design and the construction of the SNS was a multi-laboratory project involving five U.S. national labs: Argonne National Laboratory, Brookhaven National Laboratory, Lawrence Berkeley National Laboratory, Los Alamos National Laboratory, and Oak Ridge National Laboratory. Figure 8.3 shows a schematic layout of the main constituent facilities of the SNS and the main contributor to the design of each part.

The acceleration is done in the linear accelerator system to attain 1-GeV beam energy. The accumulator is a storage ring where the pulse of proton is compressed from 1 m sec pulse to 1 micro sec pulse. The bending magnets of the SNS's storage ring are shown in Figure 8.4. Then the neutrons go through the moderator to slow down and reach thermal equilibrium. This moderation process takes some 10 to 20 microseconds. Therefore, the pulse length of the incoming proton beam from the accumulator should be much less than 10 microseconds. The resulting intense pulsed neutron beam is then provided for different scientific research and industrial developments.

Figure 8.2 An aerial view of the SNS (courtesy of the Oak Ridge National Laboratory).

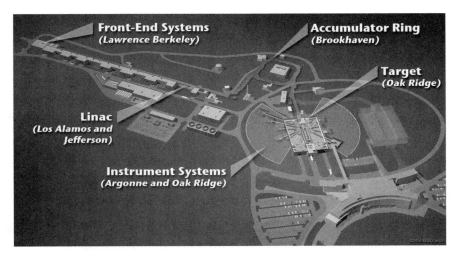

Figure 8.3 Layout of the main constituent facilities of the SNS (courtesy of the Oak Ridge National Laboratory).

8.3 Plasma-Based Accelerators

Particle accelerators discussed so far have relied on the force of RF electric fields to accelerate electrons and other particles. However, these RF accelerators, es-

Figure 8.4 Magnets in the storage ring at the SNS (courtesy of the Oak Ridge National Laboratory).

pecially colliders for high-energy physics studies, have been pushing against an upper limit for field strength of tens of MeV per meter. At higher energies, the RF electric fields can break down and could even generate enough heat to melt accelerator components if not sufficiently cooled. So the only way to get more energy from RF accelerators is to build longer, more expensive ones. Fortunately, accelerator physicists have been working hard to find alternative mechanisms for acceleration. One alternative technique is the plasma wake field acceleration. Experiments indicate that plasma wakefield machines could generate tens of GeV per meter—as much as 1,000 times more acceleration potential per length of accelerator than the conventional RF field acceleration [12–14].

8.3.1 Basic Concept of PWFA

Plasma simply can be looked at as a soup of ionized gas in which electrons are stripped from atoms. It can be generated by subjecting a diluted gas to heat, a powerful laser beam, or an electron beam. Similar to the wakes commonly seen behind moving boats, speeding electrons or a laser pulse can create a charge wake in a sea of ionized gas, or plasma. Since the atoms are already ionized in a plasma, the limit of breakdown and ionization is not of concern for charged particles accelerated in plasma.

In a plasma accelerator an intense particle beam is used to drive a longitudinal wave, which trails the driving beam at close to the speed of light. The electric fields formed between the peaks and troughs of these plasma waves can be orders of magnitude larger than those used in conventional RF accelerators.

Let us consider a case of a plasma wake generated by the passage of a near-light speed electron bunch as it is simpler in concept as well in implementation. Let us consider Figure 8.5 [15]. In (a) we show a plasma, made of positive ions and free electrons, before an electron bunch enters. As the electron bunch enters the plasma in (b), it repels the free electrons from its path, and attracts the positive ions. The moving electron bunch leaves a wake of positive ions behind it as it passes. Now we see in (c) that the displaced free electrons are attracted to the mass of positive ions behind the electron bunch. In (d), one can visualize how the free electrons in their new position can propel electrons in the trailing part of the bunch and accelerate them.

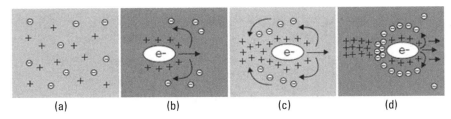

Figure 8.5 Simplified illustration of the basic concept of plasma wakefield acceleration [15].

The above effect can be demonstrated in two routines. In the single-bunch routine, the bunch is propelled by its own wake. The head of the bunch creates the plasma and drives a wake of charge. The wake is driven out of the path of the incoming electrons, creating a charge imbalance that pulls it crashing back in behind the passing electron bunch. This effect produces a strong field that accelerates particles in the back of the bunch. The system effectively operates as a transformer, where the energy from the particles in the head is transferred to those in the back, through the plasma wake. An energy gain of more than 42 GeV is achieved in a plasma wakefield accelerator of 85 cm length, driven by a 42 GeV electron beam at the SLAC Stanford Linear Accelerator Laboratory [16].

The physics is similar if there are two bunches rather than one. Here, the first bunch creates the wake (called the "drive" bunch) and a trailing bunch (called the "witness" bunch) that would get accelerated by the wake of the drive bunch. It should be pointed out that the process of energy transformation that takes place in the plasma wake field accelerator is consistent with the principle of conservation of energy. The process is based on taking energy from a large number of electrons in the drive bunch and using it to accelerate a smaller number of electrons in the witness bunch to a much higher energy.

Plasma wakefield acceleration is one of the promising approaches to advancing accelerator technology. For this reason, several national laboratories and universities are investigating and developing this new technology. At SLAC National Accelerator Laboratory, a specialized test facility is being established. The Facility for Advanced Accelerator Experimental Tests (FACET) [15] uses two-thirds of the famous two-mile-long linear accelerator at SLAC.

8.4 Dielectric Wall Accelerators

A compact *dielectric wall accelerator* (DWA) with field gradient up to 100 MV/m is being developed by a group of scientist at the Lawrence Livermore National Laboratory, in California [17–19]. It uses fast switched high-voltage transmission lines to generate pulsed electric fields on the inside of a high gradient insulation acceleration tube. The goal is to be able to accelerate proton bunches for use in cancer therapy treatment [20, 21].

8.4.1 Basic Concept of DWA

As its name implies, the DWA is an accelerator where the beam pipe, or the inside of the accelerator, is made of a dielectric. This dielectric hosts a high-gradient tangential electric field that is parallel to the axis of the accelerator tube. Hence, it can exert a force on a charged particle, such as a proton, and accelerate it. Vacuum insulators composed of alternating layers of metal and dielectric, known as *high-gradient insulators* (HGIs) are used as the dielectric wall of the accelerator. It withstands electric fields higher than conventional insulators.

Now, we need an arrangement to supply the dielectric wall with the tangential electric field that is used for acceleration. A pulse forming line, called *Blumlein*, provides this. It is basically a transmission line that is charged to a certain voltage and then discharged in a controlled manner. To discharge the transmission line arrangement, we need a switch that can close fast. The ideal closing switch for this application would be capable of high gradient operation with a very long lifetime and a low resistance when conducting. A good switch has been found to be silicon carbide (SiC) photoconductive switches [17]. A laser light is directed into the SiC switch to activate it to close. At the closing of the switch, the line starts to discharge, marking the beginning of a pulse. A pulsed electric field is formed at the inside of the accelerating tube. An accelerator can be fabricated by stacking many of the Blumleins that are terminated on one side with the SiC switches and on the other side with the HGI dielectric (Figure 8.6).

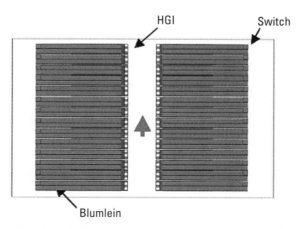

Figure 8.6 A stack made of the Blumlein terminated at SiC switches and the HGI [17].

By controlling the timing of the closing switches in the pulse forming lines, a traveling tangential electrical field can be applied to the inside dielectric wall. The speed of this field can be adjusted to keep pace with a co-moving charged particle, such as a proton, so that the particle is continually accelerated through the structure. Figure 8.7 [20] shows a simplified schematic of a proton accelerator based on the DWA.

One promising application for the DWA is to be able to dramatically reduce the size and cost of proton therapy machines (see Figures 8.8 and 8.9). The accelerator is expected to have an average acceleration gradient in the range of 100 MeV/m, and potentially can be fitted in a single treatment room [19–21]. The reader may recall from the coverage of proton therapy machines in Section 7.3 that these proton machine are based on cyclotrons or synchrotrons, where both circular machines typically have large footprints, and hence they are

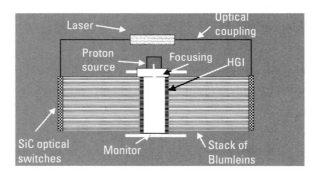

Figure 8.7 A simplified schematic of a proton accelerator based on the DWA [20].

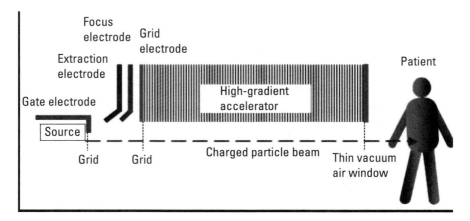

Figure 8.8 Schematic of the compact high-gradient DWA proton accelerator [17].

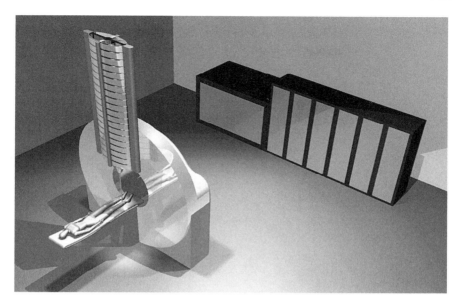

Figure 8.9 Artist's rendition of a possible proton therapy system based on the DWA [17].

relatively expensive to build. The ultimate objective of developing the DWA is to provide proton machines that can fit in a local clinic to provide proton therapy and make them more affordable.

Lawrence Livermore National Laboratory in California has a Cooperative Research and Development Agreement (CRADA) with Compact Particle Acceleration Corporation (CPAC), a subsidiary of Accuray, a radiation therapy company in the United States, to develop a compact proton DWA for radiation therapy.

8.5 Concluding Remarks

In this brief chapter, we reviewed some of the novel acceleration methods and their current and potential applications. Some of the recent developments have been put to use already, such as the neutron spallation sources and free-electron lasers. Conversely, other innovative acceleration ideas, such as the plasma wakefield acceleration and the dielectric-wall accelerators are not ready yet for implementation and require further research to overcome some of the current challenges. It is hoped that the accelerator researchers will continue pushing these advancement further, making these accelerators smaller and more affordable. This should open the door for more applications than we can envision at this time.

References

[1] Pellegrini, C., and J. Stohr, "X-Ray Free Electron Lasers: Principles, Properties and Applications," SLAC-PUB-9601, 2002, www.slac.stanford.edu/cgi-wrap/getdoc/slac-pub-9601.pdf.

[2] Arthur, J., et al., "Linac Coherent Light Source (LCLS) Conceptual Design Report," SLAC Report SLAC-R-593, April 2002, http://www.slac.stanford.edu/cgi-wrap/getdoc/slac-r-593.pdf.

[3] Altarelli, M., et al. (eds.), "XFEL: The European X-Ray Free-Electron Laser, Technical Design Report," DESY Report DESY 2006-097, 2006, http://xfel.desy.de/localfsExplorer_read?currentPath=/afs/desy.de/group/xfel/wof/EPT/TDR/XFEL-TDR-final.pdf.

[4] Yamano, Y., et al., "Design of XFEL Facility in Harima," *Proc. EPAC08*, Genoa, Italy, June 23–27, 2008, pp.466–468, http://accelconf.web.cern.ch/Accelconf/e08/papers/mopd010.pdf.

[5] Shetty, N., et al., "Interaction of High Energy Proton Beam in a Spallation Target," 2011, http://www.inbk.rwth-aachen.de/publikationen/Shetty_Nabbi_High_Energy_Proton_Beam_2011.pdf.

[6] Holtkamp, N., "The SNS Linac and Storage Ring: Challenges and Progress Towards Meeting Them," *Proc. EPAC2002*, Paris, France, Jun. 3–7, 2002, pp. 164–168.

[7] Oak Ridge National Laboratory Neutron Sciences, Spallation Neutron Source web site, http://www.sns.gov/facilities/SNS/, last accessed July 2011.

[8] ISIS web site, http://www.isis.rl.ac.uk/about-isis/aboutisis.html.

[9] Paul Scherrer Institute, SINQ: The Swiss Spallation Neutron Source web site, http://sinq.web.psi.ch.

[10] Japan Proton Accelerator Research Complex (J-Parc) web site, http://j-parc.jp/index-e.html.

[11] European Spallation Source web site, http://ess-scandinavia.eu.

[12] Esarey, E., et al., "Overview of Plasma-Based Accelerator Concepts," *IEEE Trans. on Plasma Science*, Vol. 24, No. 2, April 1996, pp. 252–288.

[13] Joshi, C., "Plasma Accelerators—Progress and Future," *Proc. Particle Accelerator Conference PAC07*, Albuquerque, New Mexico, June 25–29, 2007, pp.3845–3849.

[14] Leemans, W. P., et al., "GeV Electron Beams from a CM-Scale Accelerator," Lawrence Berkeley National Laboratory, LBNL Paper LBNL-60105, 2006, http://escholarship.org/uc/item/79p4v28z.

[15] SLAC National Accelerator Laboratory FACET User Facility web site http://facet.slac.stanford.edu/PlasmaWakefield.asp, last accessed July 2011.

[16] Blumenfeld, I., et al., "Energy Doubling of 42 GeV Electrons in a Meter-Scale Plasma Wakefield Accelerator," *Nature*, 445, February 15, 2007, 741–744.

[17] Caporaso, G. J., et al., "High Gradient Induction Accelerator," *Proc. Particle Accelerator Conference PAC07*, Albuquerque, New Mexico, June 25–29, 2007, pp.857–861.

[18] Caporaso, G. J., et al. Sequentially pulsed traveling wave accelerator. US Patent 7,576,499 B2, issued August 18, 2009.

[19] Caporaso, G. J., et al. Compact accelerator. US Patent Application No. 2005/0184686 A1, published August 25, 2005.

[20] Mackie, T. R., "Dielectric Wall Accelerator (DWA) and Distal Edge Tracking Proton Delivery System," *American Association of Physicists in Medicine (AAPM) Annual Meeting*, September 2010, http://www.aapm.org/meetings/amos2/pdf/34-8126-60029-994.pdf.

[21] Caporaso, G. J., et al. Compact accelerator for medical therapy, US Patent 7,7,10,051 B2, issued May 4, 2010.

[22] Chen, Y-J, and A. C. Paul, "Compact Proton Accelerator for Cancer Therapy," *Proc. Particle Accelerator Conference PAC07*, Albuquerque, New Mexico, June 25–29, 2007, pp.1787–1789.

Glossary

Absorbed dose   Average energy of ionizing radiation absorbed by irradiated objects

Accelerator   A device that produces beams of high-energy charged particles such as electrons or protons.

Betatron   A circular machine used to accelerate charged particles (normally electrons) to high energies.

Brachytherapy   Radiation therapy from radioactive sources inside the body.

Bragg peak   The point at which heavily charged particles such as protons or ions deposit most of their energy in a matter

Bremsstrahlung   The production of X-ray by slowing down energetic electron beam in a specific component called X-ray target.

CERN   European Organization for Nuclear Research, an accelerator laboratory in Switzerland.

Chemotherapy   Treatment with anticancer drugs.

Cobalt 60   A radioactive substance, one of cobalt isotopes used in medical and industrial applications.

Collider   A large accelerator for colliding beams head-on

Cross-linking   Linking polymer chains together to promote a difference in the polymers' physical properties.

CRT   Conformal radiation therapy. Radiation that is shaped, or "conformed," to the shape of a tumor in all three dimensions.

CT   Computed tomography. CT scan (also known as a CAT scan) is an X-ray diagnostic procedure.

Cyclotron   A circular machine used to accelerate charged particles (normally protons) to high energies.

CW Linacs   Continuous-wave linacs

Depth Dose   Absorbed dose at a specified depth inside an irradiated object.

Diffraction Patterns   The interference pattern that results when a wave undergoes diffraction when passing through a lattice of a crystal providing information about the structure of the material.

DWA   Dielectric wall accelerator

Electrostatic accelerator   A linear machine used to accelerate charged particles to high energies using DC voltages.

eV   Electron-volt. A unit of energy equal to that of an electron accelerated by a one volt potential.

FEL   Free electron laser

EB Processing   Processing materials using electron beams

Electromagnetic waves   waves characterized by variations in electric and magnetic fields. Depending on their frequency they encompass radio waves, microwave, heat, visible light, and X-rays.

Electron emission   The liberation of electrons from a surface into the surrounding space.

Electron gun   A structure that produces electrons normally by applying heat or a laser beam.

Fractionation   Splitting of a radiation therapy dose into number of sessions given over several weeks.

Gamma Rays   Electromagnetic radiation rays that come from a radioactive source such as cobalt-60.

Gantry   The rotating part of the radiation therapy machine for delivering the treatment beams from different angles around the cancer patient during treatment.

Gray (Gy)   A unit for the absorbed radiation dose. One Gray is equivalent to energy of one joule imparted by a mass of matter of one kilogram. It equals 100 rads.

GeV   A unit of energy equal to that of an electron accelerated by 10^9 volt potential.

GHz   Gigahertz. A unit of frequency equal to 10^9 cycles per second.

Hadron   An elementary particle of heavy mass such as protons or neutrons.

IMRT   Intensity modulated radiation therapy

IGRT   Image-guided radiation therapy

Ionizing radiation   Radiation of sufficient energy to displace electrons from the atoms of matter to produce ions.

Ion implantation   Altering the characteristics of a semiconductor wafer by implanting specific ions using an accelerator.

IORT   Intraoperative radiation therapy. A type of external RT using an electron beam to deliver a single large dose to a tumor bed at the time of surgery.

KeV   Kilo-electron-volt. A unit of energy equal that of an electron accelerated by a thousand volt potential.

Klystron   An RF source. A high-power amplifier in the microwave frequency range

LCLS   Linac Coherent Light Source. Free electron X-ray laser at the SLAC National Accelerator Laboratory, California.

LEP   Large Electron Positron. A circular collider of electrons and positrons at CERN.

LHC   Large Hadron Collider

Linac   Linear accelerator.

Luminosity   An index of performance for a collider concerning the rate of collisions.

Magnetron   An RF source. A high-power oscillator in the microwave frequency range

MeV   Mega-electron-volt. A unit of energy equal that of an electron accelerated by a million volt potential.

MHz   Megahertz. A unit of frequency equal a million cycles per second.

Microwave   Electromagnetic waves in the frequency range from 300 MHz to 300 GHz.

Modulator   A device for supplying high-current and high-voltage pulses to the high-power RF source.

MW   Megawatt. A unit of power; a million watts.

NDT   Nondestructive testing

PET   Positron emission tomography. A nuclear medicine imaging procedure employing radioactive isotopes that decay by emitting a positron from the nucleus.

Photon   A quantum (energy packet) of electromagnetic radiation. X-rays and gamma rays are photon radiation.

Positron   The positively charged particle that is the antiparticle (counterpart) of the electron

PWFA   Plasma wakefield acceleration

Proton   A positively charged particle of an atom. Its mass is 1836 times that of an electron.

Rad   Radiation absorbed dose, a measure of the amount of radiation absorbed by tissues (100 rad = 1 Gray).

Radiation   Energy carried by waves or a stream of particles. Visible light, X-rays, electron beam, and a proton beam are all examples of radiation.

Resonant mode   A mode of oscillation in a resonant structure (such as a cavity) characterized by a specific frequency corresponding to a certain field configuration.

RF   Radio frequency. It is the frequency of oscillation of electromagnetic waves in the range of about 3 kHz to 30 GHz. In the accelerator community, the term RF is commonly used as a synonym to microwave. They are both used interchangeably in this book.

RT   Radiation therapy. The use of ionizing radiation to treat cancer.

Simulation   A procedure using a special X-ray imaging device to plan radiation treatment. The area to be treated is located precisely and marked for treatment.

SLAC   Stanford Linear Accelerator Center, Menlo Park, California

SNS   Spallation Neutron Source, Oak Ridge, Tennessee

SW linac   Standing-wave linear accelerator

SLC   SLAC linear collider

Synchrotron   A circular accelerator

Synchrotron radiation   Photon beam resulting from deflecting charged particles such as electrons.

Undulator A periodic magnetic structure in which interference significantly shapes the synchrotron radiation produced.

Van de Graff A type of electrostatic accelerators

Wiggler A periodic magnetic structure for producing synchrotron radiation (the magnetic field of a wiggler is stronger than that of an undulator)

X-rays High-energy, ionizing, electromagnetic radiation.

X-ray diffraction A technique used in studying the crystallographic structure and physical properties of materials.

About the Author

Samy Hanna is the principal of Microwave Innovative Accelerators (MINA), a consulting company in the area of RF engineering. He has over 25 years of engineering experience in education, research, and industry. He was previously a principal engineer and a senior manager at Siemens Medical, an engineering physicist at SLAC-Stanford University, and an associate professor of electrical engineering at Polytechnic University in New York. He is a licensed professional engineer in the State of California and the inventor of seven U.S. patents. Dr. Hanna earned his graduate degrees from Caltech and Purdue University.

Index

Absorbed dose, 93
Accelerating cavities, 29
Accelerator-based neutron sources, 176–78
Accelerator-based radiation therapy, 97–98
Accelerators
    circular, 6–8
    development history, 2–10
    dielectric wall, 173, 180–83
    early development progression, 3
    electrostatic, 3, 4–5
    large, 149–69
    manufacturing techniques, 61–88
    market for, 11–13
    in our lives, 1–2
    plasma-based, 178–80
    recent developments/trends, 173–83
    RF linear, 3, 8–10
    two-plate, 16
    virtual, 17
    waveguide, 34
    See also Linear accelerators (linacs)
Acid rain, reduction of, 133–34
Adaptive radiation therapy (ART), 108–9
Automated conditioning system, 76–78
Auxiliary cavities, 30

Bake-out. *See* Thermal outgasing
Bead-pull technique QA measure, 82–84
    defined, 82
    electric field distribution, 83
    schematic, 83
    setup, 83
Beamlines, 157, 158, 159
Beam loading, 136
Beam tests, 78–79
Betatrons, 3
    concept illustration, 7
    defined, 7
    schematic, 8
Blushing, 68, 69
Brachytherapy, 11, 95
Bragg peaks, 164, 165
Brazing, 66–70
    alloys, 68, 69–70
    blushing, 68, 69
    conditions for, 67–68
    cycle, 67
    fillers, 66, 67
    furnace, 68, 70
    See also Manufacturing
Bremsstrahlung, 38, 39
Buncher cavity, 53
Bunching mechanism, 34–35
Buyer/user guidelines, 86–88

Cancer particle therapy, 163–69
Cancer radiation therapy. *See* Radiation therapy (RT)

Cargo inspection
 dual-energy CT imaging, 141–42
 dual-energy X-ray systems, 141
 recent advances in, 140–42
 scanning units, 138–40
 X-ray computed tomography, 140–41
 X-ray detector advancement, 142
Catcher cavity, 53
Cathode
 defined, 36
 magnetron, 51
 overheating, 76
Cavities
 accelerating, 29
 auxiliary, 30
 buncher, 53
 catcher, 53
 coupled, 21–27, 30
 illustrated, 18
 magnetic properties, 19
 main, 29
 quality factor (Q), 20–21
 superconducting, 155
Cavity machining, 65
Cell machining QC, 81–82
Chemical cleaning, 65–66
Circular accelerators
 betatron, 7–8
 cyclotron, 6–7
 development of, 6–8
 particle therapy (PT), 166–69
 synchrotron, 8
 See also Accelerators
Circular colliders
 defined, 149–50
 Large Electron Positron (LEP), 152–53
 Large Hadron Collider (LHC), 153–54
 synchrotron radiation, 151
 See also Colliders
Circular waveguides, 58
Circulators
 defined, 56
 four-port, 57
 power flow, 56
 three-port, 56
Cobalt teletherapy
 defined, 95–96
 illustrated, 96
 penumbra, 96

Cockcroft-Walton accelerator, 4
Colliders, 149
 circular, 149–50, 151–54
 linear, 149, 150–51
Computed microtomography (CMT), 163
Conformal radiation therapy (CRT), 104–9
 adaptive radiation therapy (ART), 108–9
 defined, 104
 intensity-modulated radiation therapy (IMRT), 106–8
 multi-leaf collimators (MLCs), 105–6
 techniques, 105
 See also Radiation therapy (RT)
Continuous-wave (CW) linacs, 137
Cooper pairs, 155
Coupled cavities, 21–27
 analysis, 22
 chain of seven, 24
 chain of three, 22
 coupling mechanisms, 25, 26
 dispersion curve, 24, 25
 electric coupling, 24, 25
 excitation frequency, 23
 field orientations for modes, 23
 illustrated, 21
 magnetic coupling, 26
 $\pi/2$ mode, 22
 π mode, 21, 22
 use of, 26–27
 zero mode, 21–27
Coupling cavities, 30
Critical magnetic field, 155
CyberKnife, 114–15
Cyclone 235, 168–69
Cyclotrons, 1, 3
 defined, 6
 frequency of rotation, 7
 operation concept, 6
 for particle therapy, 167–68

Dielectric loss, 48
Dielectric wall accelerators (DWA), 173, 180–83
 concept, 181–83
 defined, 180
 high-gradient insulators (HGIs), 181
Diffusion bonding
 defined, 70
 good, 70–71

parameters, 71
  See also Manufacturing
Disk-loaded traveling-wave (TW) linacs, 33
Dispersion curve
  defined, 24
  illustrated, 25
  standing-wave linacs (SW), 29, 30
  See also Coupled cavities
Dose(s)
  absorbed, 93
  equivalent, 93–94
  for food irradiation, 129–30
  modulation, 108
  response curves, 102
Drift-tube accelerator, 8–10
Dual-energy CT imaging, 141–42
Dual-energy X-ray systems, 141

Electrical conductivity, 63
Electric coupling
  defined, 24
  mechanisms of coupling, 25
Electric field, 17, 19
  amplitude distribution, 83
  magnetron, 52
Electron beam (EB) sterilization, 12
Electron gun
  activation, 74–76
  cathode, 36
  composition, 36
  defined, 35
  emitting surface coating, 37
  focusing electrodes, 36
  impregnation, 36–37
  operation, 35–37
  schematic, 36
  simulated emission plots, 75
Electronic portal imaging devices (EPIDs)
  defined, 110
  flat-panel, 110–11
  TV-camera-based, 110, 111
  See also Image-guided radiation
    therapy (IGRT)
Electrons
  collisions, 94
  secondary emission yield, 64
  weighting factor, 94
Electron therapy, 97
Electron-volt (eV), 17
Electrostatic accelerators, 3

Cockcroft-Walton, 4
development of, 4–5
maximum energy limitation, 5
Van de Graff, 4–5
Energy deposition
  electron beam, 123
  normalized, 122
Energy Selection System (ESS), 167
Environmental applications
  EB treatment of flue gases, 133–34
  wastewater treatment, 131–33
European Organization for Nuclear Research
    (CERN), 151
Excitation frequency, 23
Exit window, 39
External radiation therapy, 11

Fiducial markers, 108
Flat-panel EPIDs, 110–11
Flue gases, EB treatment of, 133–34
Focusing electrodes, 36
Food irradiation, 129–31
  applications, 130
  defined, 129
  food acceptance, 130–31
  pathogen-killing mechanism, 130–31
  required doses for, 129–30
Fractionation, dose, 101–2
Free-electron laser (FEL), 173, 174
  defined, 174
  light generation, 175
  linac coherent light source (LCLS), 175
  operation illustration, 174

Gamma Knife
  concept illustration, 116
  defined, 115
  illustrated, 116
Gamma rays
  defined, 93
  weighting factor, 94
Gemstone irradiation, 126
Gun activation
  defined, 74
  incomplete, 75
  overheating the cathode, 76

Hadron therapy, 164
Head-on collisions, 149

High-gradient insulators (HGIs), 181
    proton therapy system, 182–83
    schematic, 182
High-power RF conditioning, 76–78

Image-guided radiation therapy (IGRT), 109–15
    defined, 109
    electronic portal imaging devices, 110–11
    machines, 109
    need for, 109
    portal films, 110
    tomotherapy, 112–14
Industrial applications, 121–45
    EB treatment of flue gases, 133–34
    environmental, 131–34
    food irradiation, 129–31
    material processing, 121
    medical product sterilization, 127–29
    nondestructive testing (NDT), 135–37
    security and inspection, 137–42
    semiconductor chip fabrication, 143–44
    wastewater treatment, 131–33
Intensity-modulated radiation therapy (IMRT), 106–8
    beamlets, 107
    defined, 106–7
    dose modulation, 108
    example, 108
Intraoperative radiation therapy (IORT), 117–18
Ion implanters, 143–44
    defined, 143
    operation concept, 143–44
    schematic, 144
Ionizing radiation, 92–93
    direct, 92
    indirect, 92–93
    parameters, 93

Klystron
    buncher cavity, 53
    catcher cavity, 53
    defined, 50
    fundamental concept, 54
    magnetron versus, 55
    modulator, 53
    operation schematic, 54
    RF driver, 52–53
    Thales' S-band, 54–55
    use of, 54–55
    velocity modulation, 54

LabVIEW, 72
Large accelerators, 149–69
    cancer particle therapy, 163–69
    facilities for high-energy physics, 149–56
    superconductivity in, 154–56
    synchrotrons sources, 157–63
Large Electron Positron (LEP), 152–53
Large Hadron Collider (LHC), 153–54
    crossing border map, 154
    defined, 153
    parameters, 155
    shut down, 153
Linac coherent light source (LCLS), 175
Linac water-cooling system, 48–49
Linear accelerators (linacs), 1
    assembly in clean room, 67
    auxiliary systems, 45–49
    basic concepts and constituents, 15–40
    beam tests and test bunkers, 78–79
    bunching mechanism, 34–35
    in cancer radiation therapy, 91–118
    clinical use of, 100–104
    configurations, 27–35
    constituents and auxiliary systems illustration, 44
    continuous-wave (CW), 137
    efficiency with nose cones, 26
    electron gun activation, 74–76
    fundamental concepts and definitions, 16–21
    high-power RF conditioning, 76–78
    manufacturing techniques, 61–88
    medical, 98–100
    RF system, 49–59
    RF vacuum window, 46–48
    as source for electron and X-ray beams, 43–45
    standing-wave (SW), 27–32
    supporting system, 43–59
    thermal outgasing, 73–74
    traveling-wave (TW), 32–34
    tuning of, 71–73
    vacuum system, 45

water-cooling system, 48–49
Linear colliders
    acceleration mechanism, 150
    defined, 149
    interaction point (IP), 150
    luminosity, 150
    Stanford Linear Collider (SLC), 150–51
    See also Circular colliders
Loma Linda University Medical Center (LLUMC) proton therapy center, 168–69
Luminosity, 150

Magnetic coupling, 26
Magnetic field, 19
    critical, 155
    magnetron, 52
Magnetron
    anode, 51
    cathode, 51
    cutaway, 53
    defined, 50
    electric field, 52
    fundamental concept, 51
    klystron versus, 55
    magnetic field, 52
    output loop, 52
    S-band, 51
    strapping, 52
    structure, 51–52
    use of, 50
Main cavities, 29
Manufacturing, 74–76
    assembly and bonding techniques, 66–71
    beam tests and test bunkers, 78–79
    brazing, 66–70
    cavity machining, 65
    chemical cleaning, 65–66
    diffusion bonding, 70–71
    electron gun activation, 74–76
    guidelines for buyers and users, 86–88
    high-power RF conditioning, 76–78
    issues and inspections, 80
    material requirements, 63–64
    potential imperfections, 81
    processes overview, 61–62
    process flow, 62
    process flow chart, 87
    quality systems, 80–86
    techniques, 61–88
    thermal outgasing, 73–74
Materials, 63
    cost, 64
    electrical conductivity, 63
    electron secondary emission yield, 64
    mechanical stiffness, 64
    requirements, 63–64
    thermal conductivity, 63
    vacuum outgasing, 63–64
Materials processing
    electron beam current and energy requirements, 122–23
    polymer EB cross-linking, 123–26
    with X-ray radiation, 126–27
    See also Industrial applications
Medical linacs
    clinical use of, 100–104
    cost of manufacturing, 99–100
    dose-rate output, 98–99
    output stability, 99
    precision of output beam, 99
    requirements, 98–100
    size and length, 99
    See also Linear accelerators (linacs)
Medical product sterilization, 127–29
Medical radioisotopes, 12
Mobetron, 117–18
Mobile scanning units, 140
Moderators, 177
Monochromators, 159
Multi-leaf collimators (MLCs), 105–6
    defined, 105
    examples, 106, 107
    leaves, 105
    schematic, 106

Neutron scattering, 176
Neutron spallation, 176–77
    defined, 176
    Spallation Neutron Source (SNS), 177–78
Nondestructive testing (NDT), 12, 135–37
    applications, 136
    defined, 135
    linacs operation, 135
    X-ray radiography, 136
Normalized energy deposition, 122
Nose cones, 26

On-axis biperiodic standing-wave
    linacs (SW), 29
Oxygen-free high-conductivity (OFHC), 20

Particle accelerators. *See* Accelerators
Particle therapy (PT), 97, 166–69
    advantages, 164–66
    Bragg peaks, 164, 165
    cancer, 163–69
    circular accelerators, 166–69
    cyclotrons for, 167–68
    ionizing radiation in, 163
    synchrotrons for, 168–69
Pathogen-killing mechanism, 130–31
Penumbra, 96
Percentage depth dose (PDD), 94–95
Plasma-based accelerators, 178–80
    concept of, 179–80
    physics, 180
    wakefield, 180
Plasma wakefield acceleration (PWFA), 173
Polymer EB cross-linking
    applications for, 123–26
    gemstone irradiation, 126
    surface curing, 123–25
    tire rubber treatment, 125–26
    wire/cable insulation, 125
Polymers, processing of, 13
Portal films, 110
Proton therapy. *See* Particle therapy (PT)
Pulse repetition frequency (PRF), 137

Quality assurance (QA), 80
    bead-pull technique measure, 82–84
    measures, 80
Quality control (QC), 80
    cell machining measure, 81–82
    measures, 80
Quality factor (Q), 20
Quality systems
    bead-pull technique QA measure, 82–84
    cell machining QC, 81–82
    QC and QA examples, 81–84
    statistical process control, 84–86
    See also Manufacturing

Radiation therapy (RT)
    accelerator-based, 97–98
    adaptive (ART), 108–9
    brachytherapy, 95
    cancer, 10–11
    clinical requirements, 100–101
    clinical use of linacs in, 100–104
    cobalt teletherapy, 95–97
    concepts and definitions, 91–95
    conformal (CRT), 104–9
    defined, 11, 91
    dose fractionation, 101–2
    electron, 97
    external beam radiotherapy, 95
    image-guided (IGRT), 109–15
    intensity-modulated (IMRT), 106–8
    intraoperative (IORT), 117–18
    ionizing, 92–93
    linac role in, 91–118
    machine configurations, 104
    medical linac requirements, 98–100
    objective of, 11
    particle, 97
    radiation types, 92
    radionuclides-based, 95–97
    rotational therapy, 103–4
    stereotactic radiosurgery (SRS), 115–16
    treatment planning and simulation, 101
    types, 11
    use of, 92
    X-ray, 97
Radionuclides-based radiation
    therapy, 95–97
Rectangular waveguides, 58
Resistive loss, 48
Resonance, 19
Resonant coupling, 30
Resonant frequency, 18
RF driver, 52–53
RF linear accelerators, 3
    development of, 8–10
    drift-tube, 8–10
    resonant cavity, 18
    Wideröe, 10
RF system, 50–59
    circulator, 56–57
    function of, 49–50
    high-power sources, 50–55
    klystron, 52–55
    magnetron, 50–52
    power transmission subsystems, 56–59
    schematic, 50

waveguide transmission, 57–59
    See also Linear accelerators (linacs)
RF vacuum window, 46–48
    defined, 46–47
    failure, 48
    heating sources, 48
    illustrated, 47
    for S-band linac, 47
RLC resonant circuits, 19
Robotic radiosurgery, 114–15
Rotational therapy, 103–4

Scanning units, 138–40
    mobile, 140
    stationary, 138–40
Security and inspection applications, 13
Security/inspection applications
    cargo inspection, 140–42
    scanning units, 138–40
Semiconductor chip fabrication, 13, 143–44
Side-coupled standing-wave linacs (SW)
    acceleration mechanism, 32
    defined, 30
    illustrated, 32
Spallation
    defined, 176
    neutron, 176–77
Spallation Neutron Source (SNS), 177–78
    acceleration mechanism, 177
    aerial view, 178
    defined, 177
    layout, 178
    magnets in storage ring, 179
Spinning target, 40
Spread-out Bragg peak (SOBP), 165–66
Standing-wave linacs (SW), 27–32
    accelerating cavities, 29
    on-axis biperiodic, 29
    configurations, 31
    coupling cavities, 30
    defined, 27
    dispersion curve, 29, 30
    π mode, 29
    π mode schematic, 28
    resonant coupling, 30
    schematic, 28
    side-coupled, 30, 32
    standing-wave patterns, 28
    synchronism, 29
    See also Linear accelerators (linacs)

Stanford Linear Collider (SLC), 150–51
Stationary scanning units, 138–40
Statistical process control (SPC)
    bake-out time, 85
    cavity frequency, 85
    control chart, 85
    defined, 84
    E-gun activation time, 86
    lower control limit (LCL), 84
    parameters, 85–86
    receiving inspection parameters, 85
    RF conditioning time, 86
    tuning parameters, 85
    upper control limit (UCL), 84
Stereotactic radiosurgery (SRS), 115–16
    defined, 115
    Gamma Knife, 115–16
Strapping, 52
Superconductivity, 154–56
Surface curing, 123–25
Synchrotron-radiation-induced-x-ray
        emission (SRIXE), 163
Synchrotrons, 1, 3
    applications, 161–63
    beamlines, 157, 158, 159
    defined, 8
    layout illustration, 158
    light, 157
    for particle therapy, 168–69
    principle illustration of, 9
    radiation, 151, 157
    radiation, transporting and filtering,
        160
    short pulse, 159
    sources, 157–63
    storage rings, 157–60
    undulators, 160–61
    wigglers, 160–61

Target design
    approaches, 39–40
    fixed versus moving, 40
    internal versus external, 39–40
    requirements, 38–39
Test bunkers
    defined, 78
    layout, 79
Thales' S-band klystron, 54–55
Thermal conductivity, 63

Thermal outgasing, 73–74
    cycle-time, 74
    defined, 73
Tire rubber treatment, 125–26
Tomotherapy, 112–14
    defined, 112
    helical delivery, 112
    image-guided positioning, 114
    schematic representation, 112
    TomoTherapy, 112, 114
    treatment delivery, 114
Transition temperature (T_c), 155
Traveling-wave (TW) linacs, 32–34
    defined, 32–33
disk-loaded, 33
    operation concept, 33
    schematic, 33
    use of, 33–34
    as waveguide accelerators, 34
Treatment couch, 103
Tuning linacs, 71–73
TV-camera-based EPIDs, 110, 111

Undulators, 160–61

Vacuum outgasing, 63–64
Vacuum system, 45–46
Van de Graff electrostatic accelerator, 4–5
Velocity modulation, 54

Wastewater treatment, 131–33
    EB interaction schemes, 133
    effectiveness, 134
Waveguides
    bends, 59
    circular, 58
    cut-through view, 59
    defined, 34
    flexible, 58
    rectangular, 58
    transmission, 57–59
Wigglers, 160–61, 162

X-ray beams
    conversion mechanism, 38
    generation, 37–39
    linac as source, 44
    target design requirements, 38–39
X-ray computed tomography, 140–41
X-ray detector advancement, 142
X-ray diffraction, 162
X-ray NDT radiography, 136
X-rays
    defined, 93
    industrial material processing with, 126–27
    weighting factor, 94
X-ray therapy, 97

Recent Titles in the Artech House Microwave Library

Active Filters for Integrated-Circuit Applications, Fred H. Irons

Advanced Techniques in RF Power Amplifier Design, Steve C. Cripps

Automated Smith Chart, Version 4.0: Software and User's Manual, Leonard M. Schwab

Behavioral Modeling of Nonlinear RF and Microwave Devices, Thomas R. Turlington

Broadband Microwave Amplifiers, Bal S. Virdee, Avtar S. Virdee, and Ben Y. Banyamin

Computer-Aided Analysis of Nonlinear Microwave Circuits, Paulo J. C. Rodrigues

Designing Bipolar Transistor Radio Frequency Integrated Circuits, Allen A. Sweet

Design of FET Frequency Multipliers and Harmonic Oscillators, Edmar Camargo

Design of Linear RF Outphasing Power Amplifiers, Xuejun Zhang, Lawrence E. Larson, and Peter M. Asbeck

Design Methodology for RF CMOS Phase Locked Loops, Carlos Quemada, Guillermo Bistué, and Iñigo Adin

Design of RF and Microwave Amplifiers and Oscillators, Second Edition, Pieter L. D. Abrie

Digital Filter Design Solutions, Jolyon M. De Freitas

Discrete Oscillator Design Linear, Nonlinear, Transient, and Noise Domains, Randall W. Rhea

Distortion in RF Power Amplifiers, Joel Vuolevi and Timo Rahkonen

EMPLAN: Electromagnetic Analysis of Printed Structures in Planarly Layered Media, Software and User's Manual, Noyan Kinayman and M. I. Aksun

An Engineer's Guide to Automated Testing of High-Speed Interfaces, José Moreira and Hubert Werkmann

Essentials of RF and Microwave Grounding, Eric Holzman

FAST: Fast Amplifier Synthesis Tool—Software and User's Guide,
   Dale D. Henkes

Feedforward Linear Power Amplifiers, Nick Pothecary

Foundations of Oscillator Circuit Design, Guillermo Gonzalez

Frequency Synthesizers: Concept to Product, Alexander Chenakin

*Fundamentals of Nonlinear Behavioral Modeling for RF and
   Microwave Design,* John Wood and David E. Root, editors

Generalized Filter Design by Computer Optimization,
   Djuradj Budimir

High-Linearity RF Amplifier Design, Peter B. Kenington

High-Speed Circuit Board Signal Integrity, Stephen C. Thierauf

Integrated Microwave Front-Ends with Avionics Applications,
   Leo G. Maloratsky

Intermodulation Distortion in Microwave and Wireless Circuits,
   José Carlos Pedro and Nuno Borges Carvalho

Introduction to Modeling HBTs, Matthias Rudolph

Introduction to RF Design Using EM Simulators, Hiroaki Kogure,
   Yoshie Kogure, and James C. Rautio

*Klystrons, Traveling Wave Tubes, Magnetrons, Crossed-Field
   Amplifiers, and Gyrotrons,* A. S. Gilmour, Jr.

Lumped Elements for RF and Microwave Circuits, Inder Bahl

Lumped Element Quadrature Hybrids, David Andrews

Microwave Circuit Modeling Using Electromagnetic Field Simulation,
   Daniel G. Swanson, Jr. and Wolfgang J. R. Hoefer

Microwave Component Mechanics, Harri Eskelinen and
   Pekka Eskelinen

*Microwave Differential Circuit Design Using Mixed-Mode
   S-Parameters,* William R. Eisenstadt, Robert Stengel, and
   Bruce M. Thompson

Microwave Engineers' Handbook, Two Volumes,
   Theodore Saad, editor

*Microwave Filters, Impedance-Matching Networks, and Coupling
   Structures,* George L. Matthaei, Leo Young, and E. M. T. Jones

Microwave Materials and Fabrication Techniques, Second Edition,
   Thomas S. Laverghetta

Microwave Materials for Wireless Applications, David B. Cruickshank

Microwave Mixers, Second Edition, Stephen A. Maas

Microwave Network Design Using the Scattering Matrix,
   Janusz A. Dobrowolski

Microwave Radio Transmission Design Guide, Second Edition,
   Trevor Manning

Microwaves and Wireless Simplified, Third Edition,
   Thomas S. Laverghetta

Modern Microwave Circuits, Noyan Kinayman and M. I. Aksun

Modern Microwave Measurements and Techniques, Second Edition,
   Thomas S. Laverghetta

Neural Networks for RF and Microwave Design, Q. J. Zhang and
   K. C. Gupta

Noise in Linear and Nonlinear Circuits, Stephen A. Maas

Nonlinear Microwave and RF Circuits, Second Edition,
   Stephen A. Maas

Q Factor Measurements Using MATLAB®, Darko Kajfez

*QMATCH: Lumped-Element Impedance Matching, Software and
   User's Guide,* Pieter L. D. Abrie

*Passive RF Component Technology: Materials, Techniques, and
   Applications,* Guoan Wang and Bo Pan, editors

Practical Analog and Digital Filter Design, Les Thede

Practical Microstrip Design and Applications, Günter Kompa

Practical RF Circuit Design for Modern Wireless Systems, Volume I: Passive Circuits and Systems, Les Besser and Rowan Gilmore

Practical RF Circuit Design for Modern Wireless Systems, Volume II: Active Circuits and Systems, Rowan Gilmore and Les Besser

Production Testing of RF and System-on-a-Chip Devices for Wireless Communications, Keith B. Schaub and Joe Kelly

Radio Frequency Integrated Circuit Design, Second Edition, John W. M. Rogers and Calvin Plett

RF Bulk Acoustic Wave Filters for Communications, Ken-ya Hashimoto

RF Design Guide: Systems, Circuits, and Equations, Peter Vizmuller

RF Linear Accelerators for Medical and Industrial Applications, Samy Hanna

RF Measurements of Die and Packages, Scott A. Wartenberg

The RF and Microwave Circuit Design Handbook, Stephen A. Maas

RF and Microwave Coupled-Line Circuits, Rajesh Mongia, Inder Bahl, and Prakash Bhartia

RF and Microwave Oscillator Design, Michal Odyniec, editor

RF Power Amplifiers for Wireless Communications, Second Edition, Steve C. Cripps

RF Systems, Components, and Circuits Handbook, Ferril A. Losee

The Six-Port Technique with Microwave and Wireless Applications, Fadhel M. Ghannouchi and Abbas Mohammadi

Solid-State Microwave High-Power Amplifiers, Franco Sechi and Marina Bujatti

Stability Analysis of Nonlinear Microwave Circuits, Almudena Suárez and Raymond Quéré

Substrate Noise Coupling in Analog/RF Circuits, Stephane Bronckers, Geert Van der Plas, Gerd Vandersteen, and Yves Rolain

System-in-Package RF Design and Applications, Michael P. Gaynor

TRAVIS 2.0: Transmission Line Visualization Software and User's Guide, Version 2.0, Robert G. Kaires and Barton T. Hickman

Understanding Microwave Heating Cavities, Tse V. Chow Ting Chan, and Howard C. Reader

For further information on these and other Artech House titles, including previously considered out-of-print books now available through our In-Print- Forever® (IPF®) program, contact:

Artech House Publishers
685 Canton Street
Norwood, MA 02062
Phone: 781-769-9750
Fax: 781-769-6334
e-mail: artech@artechhouse.com

Artech House Books
16 Sussex Street
London SW1V 4RW UK
Phone: +44 (0)20 7596 8750
Fax: +44 (0)20 7630 0166
e-mail: artech-uk@artechhouse.com

Find us on the World Wide Web at: www.artechhouse.com